Keys to Preparing for College

CAROL CARTER
JOYCE BISHOP
SARAH LYMAN KRAVITS
with LESA HADLEY

Upper Saddle River, New Jersey 07458

Library of Congress Cataloging-in-Publication Data

Carter, Carol.
Keys to preparing for college / Carol Carter, Joyce Bishop, Sarah Lyman Kravits with Lesa Hadley.— 1st ed.
p. cm.
Includes bibliographical references and index.
ISBN 0-13-030806-4
1. College student orientation—United States—Handbooks, manuals, etc. 2. Universities and colleges—United States—Admission—Handbooks, manuals, etc. 3. Study skills—United States—Handbooks, manuals, etc. I. Bishop, Joyce (Joyce L.), 1950- II. Kravits, Sarah Lyman. III. Title.

LB2343.32.C369 2001
378.1'98—dc21 00-040644

Acquisitions Editor: Sande Johnson
Assistant Editor: Michelle Williams
Production Editor: Holcomb Hathaway
Director of Manufacturing and Production: Bruce Johnson
Managing Editor: Mary Carnis
Manufacturing Manager: Ed O'Dougherty
Art Director: Marianne Frasco
Marketing Manager: Jeff McIlroy
Marketing Assistant: Barbara Rosenberg
Cover Design: Joe Sengotta
Cover Art: Jude Maceren, Images.Com/SIS
Composition: Aerocraft Charter Art Service
Printing and Binding: The Banta Company

Prentice-Hall International (UK) Limited, *London*
Prentice-Hall of Australia Pty. Limited, *Sydney*
Prentice-Hall Canada Inc., *Toronto*
Prentice-Hall Hispanoamericana, S.A., *Mexico*
Prentice-Hall of India Private Limited, *New Delhi*
Prentice-Hall of Japan, Inc., *Tokyo*
Pearson Education Singapore Pte. Ltd.
Editora Prentice-Hall do Brasil, Ltda., *Rio de Janeiro*

10 9 8 7 6 5 4 3 2 1
ISBN 0-13-030806-4

Contents

Preface

AS YOUR AUTHORS, we have talked to students across the country. We've learned that you are concerned about your future; you want to learn how to make decisions about college, majors, degrees, and careers; you want your education to serve a purpose; and you want honest and direct guidance on how to achieve your goals. We designed the features of *Keys to Preparing for College* based on what you have told us about your needs.

THE CONTENTS OF THE PACKAGE: WHAT'S INCLUDED

We chose the topics in this book based on what you need to make the most of your educational experience. You need a strong sense of self, goals, and decision making in order to discover and pursue the best course of study. You need guidance on the basic steps it takes to move from high school to college, and you need study skills to take in and retain what you learn both in and out of class. You need personal management skills—such as time management, communication, and leadership skills—to make the most of who you are.

The distinguishing characteristics and sections of this book are designed to make your life easier by helping you take in and understand the material you read.

Lifelong Learning

The ideas and strategies you learn that will help you succeed in school are the same ones that will bring you success in your career and in your personal life. Therefore, this book focuses on success strategies as they apply to school, work, and life, not just to the classroom.

Thinking Skills

Being able to remember facts and figures won't do you much good at school or beyond unless you can put that information to work through clear and competent thinking. This book has a chapter on critical thinking and decision making that will help you explore your mind's actions and will guide you through the process of making clear decisions.

Skill-Building Exercises

Today's graduates need to be effective thinkers, team players, writers, and strategic planners. The exercises at the end of the chapters will encourage you to develop these valuable future career skills and to apply thinking processes to any topic or situation.

User-Friendliness

The following features will make your life easier in small but significant ways:

- **Exercises.** The exercises are together at the ends of the chapters, conveniently located for you.
- **Definitions.** Selected words are defined in the margins of the text.
- **Layout and style.** The book is readable, friendly, and accessible, with color added to enhance the presentation.
- **Long-term usefulness.** This text can move with you both through high school and college until your college graduation day. It provides useful information that can be applied to school, work, and life.

TAKE ACTION: READ

You are responsible for your education, your growth, your knowledge, and your future. The best we can do is offer some great suggestions, strategies, ideas, and systems that can be helpful. Ultimately, it is up to you to use whatever fits your particular self, with all your particular situations, needs, and wants, and make it your own. You've made a terrific start by choosing to think about pursuing a higher education—take advantage of all it has to give you.

Acknowledgments

This book has come about through a group effort. We would like to take this opportunity to acknowledge the people who have made it happen. Many thanks to:

- Our student editors: Aziza Davis and Michael Jackson.
- Our student reviewers: Sandi Armitage, Marisa Connell, Jennifer Moe, and Alex Toth.
- Our reviewers: Glenda Belote, Florida International University; John Bennett; Jr., University of Connecticut; Ann Bingham-Newman, California State University–Los Angeles; Mary Bixby, University of Missouri–Columbia; Barbara Blandford, Education Enhancement Center at Lawrenceville, New Jersey; Jerry Bouchie, St. Cloud State University; Mona Casady, SW Missouri State University; Janet Cutshall, Sussex County Community College; Valerie DeAngelis, Miami-Dade Community College; Rita Delude, N.H. Community Technical College; Judy Elsley, Weber State University, Utah; Sue Halter, Delgado Community College; Suzy Hampton, University of Montana; Maureen Hurley, University of Missouri–Kansas City; Karen Iverson, Heald Colleges; Stephen Jones, Drexell University; Kathryn Kelly, St. Cloud State University; Nancy Kosmicke, Mesa State College, Colorado; Frank T. Lyman, Jr., University of Maryland; Barnette Miller Moore, Indian River Community College, Florida; Rebecca Munro, Gonzaga University, Washington; Virginia Phares, DeVry of Atlanta, Georgia; Brenda Prinzavalli, Beloit College, Wisconsin; Ann Russell, Meadowdale High School, Seattle, Washington; Jacqueline Carolyn Smith, University of Southern Indiana; Joan Stottlemyer, Carroll College, Montana; Thomas Tyson, SUNY Stony Brook; Rose Wassman, DeAnza College; and Michelle G. Wolf, Florida Southern College.
- The PRE 100 instructors at Baltimore City Community College, Liberty Campus; especially President Dr. Jim Tschechtelin; Coordinator Jim Coleman; Rita Lenkin Hawkins; Sonia Lynch; Jack Taylor; and Peggy Winfield. Thanks also to Prentice Hall representative Alice Barr.
- The instructors at DeVry, especially Susan Chin and Carol Ozee.
- The instructors at Suffolk Community College.
- Prentice Hall representative Carol Abolafia.

- Our editorial consultant Rich Bucher, professor of sociology at Baltimore City Community College.
- R. Andrew Burt, leadership trainer.
- Dr. Frank T. Lyman, inventor of the Thinktrix system.

Finally, for their ideas, opinions, and stories, we would like to thank all of the students and professors with whom we work. Joyce in particular would like to thank the thousands of students who have allowed her the privilege of sharing part of their journey through school. We appreciate that, through reading this book, you give us the opportunity to learn and discover with you.

Keys to Preparing for College

Part I

Keys to Choosing the Right College

Chapter 1

Setting Goals Toward Success

Chapter 2

Thinking Critically About College Decisions

Chapter 3

Exploring Careers and Majors

Chapter 4

Choosing a College that Fits Your Needs

R E A C H

1 Setting Goals Toward Success

Since you were small, family and friends have probably asked you that old question, "What do you want to be when you grow up?" Perhaps you had an answer for that question, but perhaps your answer changed everytime someone asked. Now that graduation from high school is nearing, you are probably hearing a different version of that question, "What are you going to do when you graduate?" Perhaps you are one of the lucky few who knows exactly what you want to achieve and the steps you will take to achieve that goal. However, you may be like many students who feel overwhelmed by all of the possibilities.

Over the next year, you will have to make many decisions that could have a huge impact on your life. Will you attend college? What kind of college will you attend? What will you major in? These are serious questions that you will need to answer. Even if you don't know the exact answer to all of these questions yet, you probably have a general idea of your interests and what you would like to do—or not do—with your life. To help you answer those questions and set a path toward success, you should begin identifying goals and prioritizing the tasks you must complete.

This chapter explains how goals can make a significant difference in your life. You will explore what values are important to you and how values relate to goals, how to create a framework for your goals, how to set long-term and short-term goals, and how to set priorities.

In this chapter, you will explore answers to the following questions:

- Why do goals matter?
- How do I set and achieve goals?
- How do I evaluate goals?
- What are my priorities?

WHY DO GOALS MATTER?

Goal
An end toward which effort is directed; an aim or intention.

Perhaps you've always dreamed about a career in nursing or veterinary medicine. What does it take to move from dreamland to reality? It takes goals! The steps one takes in setting a **goal** to be a nurse or veterinarian or to be in any other profession help move that dream into a reality because they force a person to explore the specific steps that need to be taken to reach that goal. When tasks that seem overwhelming are broken into subtasks or short-term goals, the goal becomes manageable and decisions become easier to make. Suddenly those pie-in-the-sky dreams become goals that you can more easily move toward.

Setting Goals Allows You to Reflect and Act on What Is Important to You

As you will discover in this chapter, one of the first steps in goal setting is to explore what it is that you value. When you are acting on your personal goals and not someone else's, life takes on more meaning. For example, older people such as parents may have wanted you to follow in their footsteps and join them in a business. You, however, would prefer to put your artistic talents to work as a graphic designer. Although both are great jobs, you should choose the path that would be meaningful to you, not necessarily to others.

Setting Goals Allows You to See the Whole Picture

Being a chemical engineer and working for an elite company may be a realistic goal for you, but how do you go from being a high school graduate to that employee? There are many steps involved in the process, but by setting that goal, you allow yourself the opportunity to explore what those steps are. In effect, you are able to map out your journey when you know the destination.

Setting Goals Makes Decision Making Easier

Goals act as a framework for the decisions you make in your life. They give you a concrete reason for doing the things you do. As you set a goal and see both the big picture and the smaller steps it takes to achieve that goal, making decisions that move you toward that goal becomes clearer, and the choices you need to make become easier. If, for instance, you know that you want to pursue a four-year education but will have a tight financial budget, it becomes easier to see that you would probably be better off purchasing a dependable, lower-priced car rather than an expensive sports car that uses a lot of gas. The more affordable car may not be your first choice, but when you have a concrete goal to follow, it becomes easier to make and follow through with the hard decisions that will help you achieve your goals.

Setting Goals Gives a Person Self-Confidence and a Positive Image

A common question that job employers ask during interviews is, "What do you want to be doing in five years?" Someone who can quickly and clearly answer this question will always have the advantage over someone who doesn't have a clear picture of his or her future. Being able to tell others what your goals are impresses others, and they will view you as someone who knows the first step toward success. And, when others view you as successful, it is easier to view yourself as successful.

If you are feeling overwhelmed and unsure about all the decisions you face in the next several years, you should consider learning how to effectively set goals.

HOW DO I SET AND ACHIEVE GOALS?

A goal can be as concrete as enrolling in college or as abstract as working to control your temper. When you set goals and work to achieve them, you engage your intelligence, abilities, time, and energy in order to move ahead. From major life decisions to the tiniest day-to-day activities, setting goals will help you define how you want to live and what you want to achieve.

Goal setting involves exploring and identifying your personal values, evaluating goals in terms of your values, defining your personal mission statement, placing your goals in long-term and short-term time frames, and linking your goals to the five life areas (personal, family, school/career, financial, and lifestyle).

Identifying My Values

Your personal **values** are the beliefs that guide your choices. Examples of values include family togetherness, a good education, caring for others, and worthwhile employment. The sum total of all your values is your value system. You demonstrate your particular value system in the priorities you set; the ways in which you communicate with others; the decisions you make

Values
Principles or qualities that one considers important, right, or good.

regarding your family, educational goals, and career choices; and even the material things with which you surround yourself.

Sources of Values

Examining the sources of your values can help you define those values, trace their origin, and question the reasons why you have adopted them. Value sources, however, aren't as important as the process of considering each value carefully to see if it makes sense to you. Some of your current values may have come from television or other media but still ring true. Some may have come from what others have taught you. Some you may have constructed from your own personal experience and opinion. You make the final decisions about what to value, regardless of the source.

Each individual value system is unique, even if many values come from other sources. Your value system is yours alone. Your responsibility is to make sure that your values are your own choice, not the choice of others. Make value choices for yourself based on what feels right for you, for your life, and for those who are touched by your life.

Evaluating Goals in Relation to My Values

Understanding your values will help you set career and personal goals because the most ideal goals help you achieve what you value. If you value spending time with your family, a related goal may include living near your parents while you are in college. A value of financial independence may generate goals, such as working while going to school and keeping credit card debt low, that reflect the value.

Goals enable you to put values into practice. When you set and pursue goals that are based on values, you demonstrate and reinforce values by taking action. The strength of those values, in turn, reinforces your goals.

You will experience a much stronger drive to achieve if you build goals around what is most important to you. For example, if a student values his cultural background, he may want to pursue a career that allows him to emphasize that value. If that student also values children, he might consider combining those values and becoming a bilingual elementary education major. When students set goals that reflect their values, the goals become easier to accomplish, and the tasks related to those goals are pleasant and fulfilling.

The opposite is also true. If a person values independence and personal freedom but sets a goal to work in a business setting in which she has to punch a time clock every day and only has two weeks of vacation time, that person will very likely be unhappy. The tasks associated with the job become meaningless and unfulfilling.

Because life changes and new experiences may bring a change in values, try to continue to evaluate your values as time goes by. Periodically evaluate the effects that having each value has on your life and see if a shift in values might suit your changing circumstances. For example, after growing up in a homogeneous community, a student who meets other students from unfamiliar backgrounds may learn a new value of living in a diverse community.

Your values will grow and develop as you do if you continue to think them through. And, as your values change, your goals may also change. To be successful at setting and achieving goals, you should continue to link your values to your goals. One way to accomplish this is to write a personal mission statement.

Defining My Personal Mission Statement

Some people go through their lives without ever really thinking about what they can do or what they want to achieve. If you choose not to set goals or explore what you want out of life, you may look back on your past with a sense of emptiness. You may not be able to articulate your accomplishments or understand why you did what you did. However, you can avoid that emptiness by periodically taking a few steps back and thinking about where you've been and where you want to be.

One helpful way to determine your general direction is to write a **personal mission statement.** Dr. Stephen Covey, author of the best-seller *The Seven Habits of Highly Effective People,* defines a mission statement as a philosophy that outlines what you want to be (character), what you want to do (contributions and achievements), and the principles by which you live. Dr. Covey compares the personal mission statement to the Constitution of the United States, a statement of principles that gives this country guidance and standards in the face of constant change.[1]

Personal mission statement
A statement of philosophy that outlines what you want to be, what you want to do, and the principles by which you live.

You may have noticed mission statements posted at many businesses. A company, like a person, needs to establish standards and principles that guide its many activities. Companies often have mission statements so that each member of the organization, from the custodian to the president, clearly understands what to strive for. If a company fails to identify its mission, a million well-intentioned employees might focus their energies in just as many different directions, creating chaos and low productivity.

Here is a mission statement from Northwest Airlines. It is displayed inside its company buildings and on the back of every employee's business card. Notice how it reinforces the company's goals of teamwork, leadership, and excellence:

> To build together the world's most preferred airline with the best people, each committed to exceeding our customers' expectations every day.

Another example is from Prentice Hall, the company that publishes this text:

> To provide the most innovative resources—books, technology, programs—to help students of all ages and stages achieve their academic and professional goals inside the classroom and out.

Successful individuals often write their own personal mission statements. These statements work much the same way that business mission statements do. They keep the individuals focused on what they value and on their goals. Here is an example of author Carol Carter's personal mission statement:

> My mission is to use my talents and abilities to help people of all ages, stages, backgrounds, and economic levels achieve their human potential through fully developing their minds and their talents. I also aim to balance work with people in my life, understanding that my family and friends are a priority above all else.

Keegan Kautzky is a high school senior who actively pursues goals that reflect his personal mission statement. Here is an example of Keegan's mission statement:

> I want to make a positive difference in the lives of others, even if only in a small way. I want education and learning to be a lifelong commitment. I want to live a life that I will never regret or look back on as unfulfilled.

Writing a mission statement is much more than an in-school exercise. It is truly for you. Thinking through your personal mission statement can help you begin to take charge of your life. It helps to put you in control instead of allowing circumstances and events to control you. If you frame your mission statement carefully so that it truly reflects your goals, it can be your guide in everything you do.

Creation of a Personal Mission Statement

If you are not sure how to start formulating your mission statement, look to your values to guide you. Define your mission statement and goals based on what is important to you.

For instance, if you value physical fitness, your mission statement might emphasize your commitment to staying in shape throughout your life. Later in the chapter, you will learn how to break these larger mission statements into long- and short-term goals.

Writing a personal mission statement doesn't have to be a monumental task, but a few key steps will help guide you as you begin creating your statement. Brainstorming the following five questions will help you start on your own personal mission statement:

"Whether you believe you can or whether you believe you can't, you are absolutely right."

1. Who are the people you admire, and what are the qualities you admire about them? Qualities might include ambition, caring, organization, wisdom, etc.
2. What do you value? Friendships? Family? Money? Physical fitness?
3. What are your strengths and talents?
4. What do you want to do or work toward during the next two to three years?
5. How would you like other people to see you? How would they describe your characteristics?

When you have finished your brainstorming list, highlight the items that are most important to you, and begin working on a three- or four-sentence statement that clearly reflects your values, talents, and goals. Use the examples in this chapter to help guide you, or use one of the online interactive mission builders found at the Franklin Covey Web site to get started. The URL for that site is **www.franklincovey.com/cgi-bin/teens/teens-msb/part01/.**

Think of your personal mission statement as a working, living document. It isn't written in stone and should change as you move from one phase of life to the next—from single person to spouse, from parent to single parent to caregiver of an older parent. Stay flexible and reevaluate your personal mission statement from time to time.

Placing My Goals in Time Frames

Everyone has the same twenty-four hours in a day, but it often doesn't feel like enough. Have you ever had a busy day flash by so quickly that it seems you accomplished nothing? Have you ever felt that way about a longer period of time, like a month or even a year? Your commitments can overwhelm you unless you decide how to use time to plan your steps toward goal achievement.

If developing a personal mission statement establishes the big picture, placing your goals within particular time frames allows you to bring individual areas of that picture into the foreground. It's a rare goal that is reached overnight. Lay out the plan by breaking a long-term goal into stages of what you will accomplish in one day, one week, one month, six months, one year, five years, ten years, even twenty years. Planning your progress step-by-step will help you maintain your efforts over the extended time period often needed to accomplish a goal. Goals fall into two categories: long-term and short-term.

Setting Long-Term Goals

Establish first the goals that have the largest scope, the long-term goals that you aim to attain over a lengthy period of time (up to a few years or more). As a student, you know what long-term goals are all about. You are in the process of setting goals about attending college and earning a degree or certificate. Becoming educated is an admirable goal that takes a good number of years to reach.

Some long-term goals are lifelong, such as a goal to continually learn more about yourself and the world around you. Others have a more definite end, such as a goal to complete a course successfully. To determine your long-term goals, think about what you want out of your educational, professional, and personal life. Here are Carol's and Keegan's long-term goal statements:

> **Carol's goals:** To accomplish my mission through writing books, giving seminars, and developing programs that create opportunities for students to learn and develop. To create a personal, professional, and family environment that allows me to manifest my abilities and duly tend to each of my responsibilities.
>
> **Keegan's goals:** My goals are to go to college and study science and languages. I want a career working with and teaching people. Traveling and immersing myself in other cultures are very important to me, and the arts will always play a large role in my life.

For example, you may establish long-term goals such as these:

- I will graduate from school and know that I have learned all that I could, whether my grade point average (GPA) shows it or not.
- I will use my current and future job experience to develop practical skills that will help me later in life.
- I will build my leadership and teamwork skills by forming positive, productive relationships with classmates, instructors, and coworkers.

Long-term goals don't have to be lifelong goals. Think about your long-term goals for the coming year. Considering what you want to accomplish in a year's time will give you clarity, focus, and a sense of what needs to take place right away. When Carol thought about her long-term goals for the coming year, she came up with the following list:

1. Develop programs to provide internships, scholarships, and other quality initiatives for students.
2. Write a book for students emphasizing an interactive, highly visual approach to learning.
3. Allow time in my personal life to eat well, run five days a week, and spend quality time with family and friends. Allow time daily for quiet reflection and spiritual devotion.

Because Keegan is at a different point in his life than Carol is, his goals for the coming year are somewhat different. His goals include the following:

1. Apply for admission at colleges that offer majors in languages, science, and education.
2. Find a way to visit a new friend in Germany, and travel around Europe for the summer.
3. Continue to spend quality time with family members and friends, experiencing life to its fullest.

In the same way that Carol's and Keegan's goals are tailored to their personalities and interests, your goals should reflect who you are. Personal mission statements and goals are as unique as each individual.

Continuing the example above, you might adopt these goals for the coming year:

1. I will earn passing grades in all my classes.
2. I will research colleges that offer degrees in the areas I'm interested in pursuing.
3. I will join two clubs and make an effort to take leadership roles in each.

Setting Short-Term Goals

When you divide your long-term goals into smaller, manageable goals that you hope to accomplish within a relatively short time, you are setting short-term goals. Short-term goals narrow your focus, helping you to maintain your progress toward your long-term goals. They are the steps that take you where you want to go. Assume you have set the three long-term goals you just read in the previous section. To stay on track toward those goals, you may want to accomplish these short-term goals in the next six months:

- I will pass Business Writing I so that I can move on to Business Writing II.
- I will meet with my counselor to discuss area colleges.
- I will attend four meetings of the journalism club.

These same goals can be broken down into even smaller parts, such as one month:

- I will complete five of the ten essays for Business Writing I.
- I will find out the entrance requirements for area colleges that interest me.
- I will write an article for next month's journalism club newsletter.

In addition to monthly goals, you may have short-term goals that extend for a week, a day, or even a couple of hours in a given day. Take as an example the task of applying for admission to colleges that interest you. Such short-term goals may include the following:

- Three weeks from now: Write cover letters, assemble packets, and mail to colleges.
- Two weeks from now: Have drafts of any writing requirements written, and make copies of other required documents.
- One week from now: Collect required documents and set up filing system.
- By the end of today: Bookmark and copy entrance requirements in all catalogs.
- By 3 P.M. today: Collect college catalogs.

As you consider your long-term and short-term goals, notice how all of your goals are linked to one another and linked to your personal mission statement. As Figure 1.1 and Table 1.1 show, your personal mission statement establishes a context for your long-term goals. Your long-term goals establish a context for the short-term goals; in turn, your short-term goals make the long-term goals seem clearer and more reachable. The whole system works to keep you on track. Table 1.1 illustrates the connections for both Carol and Keegan.

Linking goals together. FIGURE 1.1

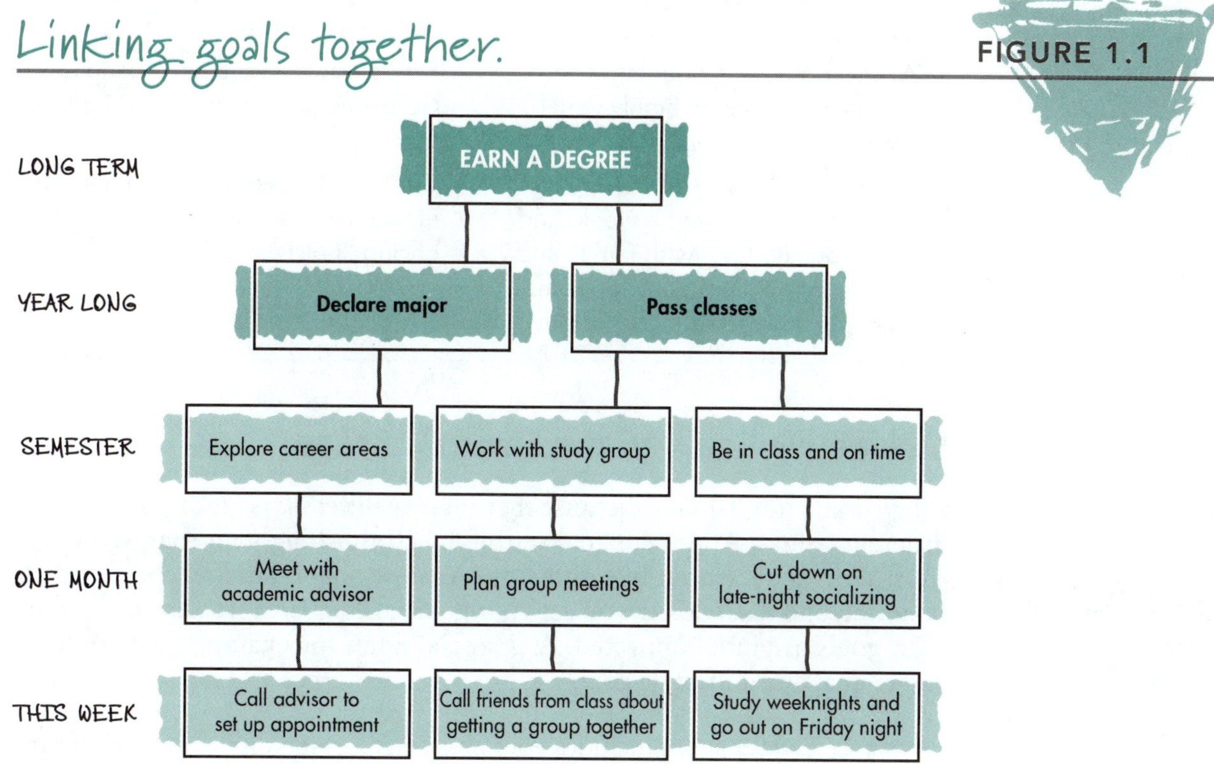

Connecting personal mission statements and goals.

	CAROL CARTER	KEEGAN KAUTZKY
Personal Mission Statement	My mission is to use my talents and abilities to help people of all ages, stages, backgrounds, and economic levels achieve their human potential through fully developing their minds and their talents. I also aim to balance work with people in my life, understanding that my family and friends are a priority above all else.	I want to make a positive difference in the lives of others, even if only in a small way. I want education and learning to be a lifelong commitment. I want to live a life that I will never regret or look back on as unfulfilled.
Long-Term Goals	To accomplish my mission through writing books, giving seminars, and developing programs that create opportunities for students to learn and develop. To create a personal, professional, and family environment that allows me to manifest my abilities and duly tend to each of my responsibilities.	My goals are to go to college and study science and languages. I want a career working with and teaching people. Traveling and immersing myself in other cultures are very important to me, and the arts will always play a large role in my life.
Short-Term Goals	1. Develop programs to provide internships, scholarships, and other quality initiatives for students. 2. Write a book for students emphasizing an interactive, highly visual approach to learning. 3. Allow time in my personal life to eat well, run five days a week, and spend quality time with family and friends. Allow time daily for quiet reflection and spiritual devotion.	1. Apply for admission at colleges that offer majors in languages, science, and education. 2. Find a way to visit a new friend in Germany, and travel around Europe for the summer. 3. Continue to spend quality time with family members and friends, experiencing life to its fullest.

Linking My Goals to the Five Life Areas

All goals are not the same because they involve different parts of your life and different values. Approach goal setting by establishing your long-term and short-term goals within five different areas: personal, family, school/career, financial, and lifestyle. As you set your goals in each area, remember that all your goals are interconnected. A financial goal, for example, will affect a career goal and a lifestyle goal.

Personal. This category includes your character, personality, physical appearance, and conduct. Do you want to gain confidence and knowledge? Develop a lean, athletic physique? Stop hanging out with people who bring you down? You can set your personal goals by taking a hard look at the difference between who you are and who you want to be.

Family. Do you want to stay single or marry? Do you want to have one or more children? Do you want to address problems with parents or change the way you relate to your family? Do you want to live near relatives or far away? The goals you set can help you build a solid, satisfying family life.

School/Career. What kind of subjects or career field do you prefer? In school, consider the classes, instructors, and class schedule. What kind of degrees or certificates are available at the college you are considering? Think about your commitment to academic excellence and whether honors and awards are important goals. Then, think about the job you want after you graduate. Consider the requirements (degrees, certificates, or tests), job duties, hours, coworkers, salary, transportation, and company size and style that might be associated with your ideal job. Do you want to become a manager, a supervisor, an independent contractor, or a business owner? How much responsibility do you want? Identify goals that can point you toward your ideal education and career.

"Even if you're on the right track, you'll get run over if you just sit there."
Will Rogers

Financial. How much money do you need to pay your bills, maintain your chosen lifestyle, and save for the future? Do you need to borrow money for school or a major purchase such as a car? Compare a realistic projection about your college financial picture to how comfortable you eventually want to be, and set goals that will help you bridge the gap. These goals will also affect the career you choose.

Lifestyle. Where do you want to live (city, suburbs, country) and in what kind of space (apartment, condominium, townhouse, single-family or multifamily house, mobile home)? What kinds of values do you want to live by and encourage in others? How do you equip yourself with the skills necessary for dealing with diverse people? With whom do you want to live (extended/immediate family, roommates, friends, no one)? What do you want to give back to your community through service or volunteer work? What do you like to do in your leisure time? Consider goals that allow you to live the way you want to live.

Setting and working toward goals can be frightening and difficult at times. Like learning a new physical task, it takes a lot of practice and repeated efforts. As long as you do all that you can to achieve a goal, you haven't failed, even if you don't achieve it completely or in the time frame you had planned. Even one step in the right direction is an achievement. For example, if you wanted to raise your course grade to a B from a D and ended up with a C, you have still accomplished something important.

Identifying Educational Goals

Education is a major part of your life right now. In order to define a context for your school goals, explore why you have decided to pursue an education. People have many reasons for attending college. You may identify with one or more of the following possible reasons:

- I want to earn a higher salary.

- I want to build marketable skills.
- My supervisor at work says that a degree will help me move ahead in my career.
- Most of my friends are going.
- I want to be a student and learn all that I can.
- It seems like the only option for me right now.
- Everyone in my family goes to college; it's expected.
- I don't feel ready to jump into the working world yet.
- I received a scholarship.
- My parents are pushing me to go to college.
- I am pregnant and need to increase my skills so I can provide for my baby.
- I want to study for a specific career.
- I don't really know.

All of these answers are legitimate, even the last one. Being honest with yourself is crucial if you want to discover who you are and what life paths make sense for you. Whatever your reasons are for considering a college degree or certificate, you are at the gateway to a journey of discovery.

It isn't easy to enroll in college, pay tuition, decide what to study, sign up for classes, gather the necessary materials, and actually get yourself to the school and into the classroom. Don't worry if you go through periods of low motivation. Remember, asking important questions gives you power to make responsible decisions that are yours and yours alone. Recharge by asking yourself, What do I want out of my life? What would I like people to say about me? What is important to me? Now and again, you may let a day get past you without making any progress, but don't let a whole life go by.

Achieving goals becomes easier when you are realistic about what is possible. Evaluating your goals will help make sure you are moving in the right direction.

HOW DO I EVALUATE GOALS?

As we set goals and work through them, we should continually be evaluating them. As our lives change, our values change. Because our value system is at the core of our goals, our goals, too, need to be adjusted.

William B. Werther, Jr., created the SMART method to evaluate goals.[2] He tells us that goals should be:

Simple Goals that are too complex do little to motivate and illuminate those who must achieve them.

Measurable Goals that can't be measured are difficult to track and will eventually be dropped.

Accountable Goals that hold an individual accountable create a sense of urgency and purpose.

Realistic Goals that are realistic have meaning, and do-able deeds make a significant contribution.

Timely Goals that have a time dimension are more action oriented.

When should you evaluate your goals? Obviously, the first time you should evaluate your goals is when you set them. Work through the SMART system to see if the goals you set are achievable. You should also evaluate your goals periodically. When circumstances in our lives change, sometimes we have to change our goals. Having a system that you can use in a consistent manner will keep you focused on both the long- and short-term goals.

When you believe you have created SMART goals, it is time to set priorities.

WHAT ARE MY PRIORITIES?

When you set a **priority,** you identify what's important at any given moment. Prioritizing helps you focus on your most important goals, even when they are difficult to achieve. If you were to pursue your goals in no particular order, you might tackle the easy ones first and leave the tough ones for later. The risk is that you might never reach for goals that are important to your success. Setting priorities helps you focus your plans on accomplishing your most important goals.

Priority
An action or intention that takes precedence in time, attention, or position.

To explore your priorities, think about your personal mission statement and look at your goals in the five life areas: personal, family, school/career, finances, and lifestyle. These five areas may not all be equally important to you right now. At this stage in your life, which two or three are most critical? Is one particular category more important than others? How would you prioritize your goals from most important to least important?

You are a unique individual, and your priorities are yours alone. What may be top priority to someone else may not mean that much to you, and vice versa. You can see this in Figure 1.2, which compares the priorities of

Two students compare priorities.

FIGURE 1.2

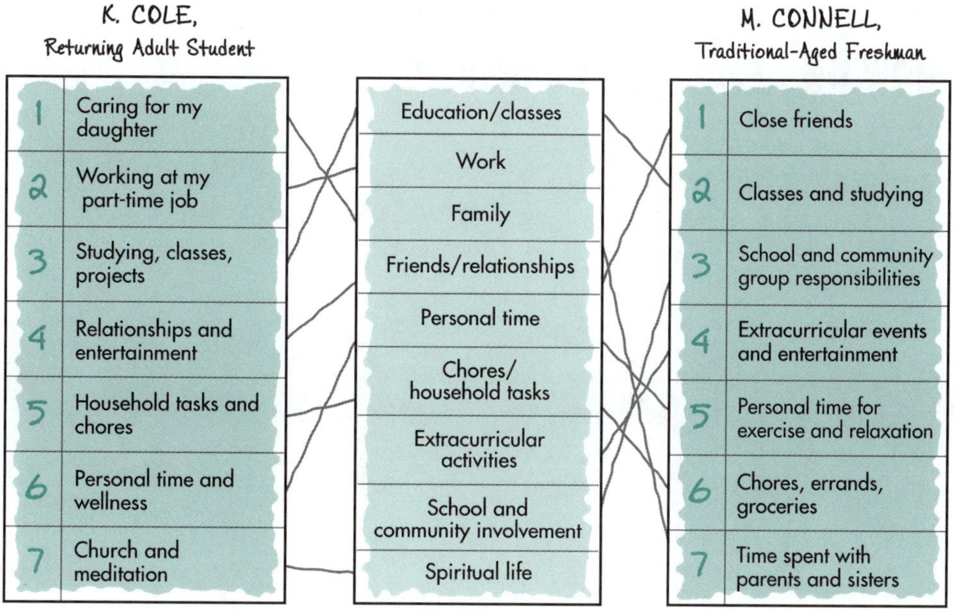

K. COLE, Returning Adult Student

1 Caring for my daughter
2 Working at my part-time job
3 Studying, classes, projects
4 Relationships and entertainment
5 Household tasks and chores
6 Personal time and wellness
7 Church and meditation

Education/classes
Work
Family
Friends/relationships
Personal time
Chores/household tasks
Extracurricular activities
School and community involvement
Spiritual life

M. CONNELL, Traditional-Aged Freshman

1 Close friends
2 Classes and studying
3 School and community group responsibilities
4 Extracurricular events and entertainment
5 Personal time for exercise and relaxation
6 Chores, errands, groceries
7 Time spent with parents and sisters

two very different students. Each student's priorities are listed in order, with the first priority at the top and the lowest priority at the bottom.

First and foremost, your priorities should reflect your personal goals. Even as you consider the needs of others, though, never lose sight of your personal goals. Be true to your goals and priorities so that you can make the most of who you are.

Setting priorities moves you closer to accomplishing specific goals. It also helps you begin planning to achieve your goals within specific time frames. And, when you have set your goals and evaluated them, you can more easily begin to think critically and make decisions that will move you toward those goals.

EXERCISES AND ACTIVITIES

My Values 1.1

Begin to explore your values by rating the following values on a scale from 1 to 4, 1 being least important to you and 4 being most important. If you have values that you don't see in the chart, list them in the blank spaces and rate them. Considering your priorities, write your top five values on a sheet of paper.

VALUE	RATING	VALUE	RATING
Knowing yourself		Mental health	
Physical health		Fitness and exercise	
Spending time with your family		Close friendships	
Helping others		Education	
Being well-paid		Being employed	
Being liked by others		Free time/vacations	
Enjoying entertainment		Time to yourself	
Spiritual/religious life		Reading	
Keeping up with the news		Staying organized	
Being financially stable		Having an intimate relationship	
Creative/artistic pursuits		Self-improvement	
Lifelong learning		Facing your fears	

My Personal Mission Statement 1.2

Using the personal mission statement examples in the chapter as a guide, consider what you want out of your life, and create your own personal mission statement. You can write it in paragraph form, in a list of long-term goals, or in the form of a think link. Take as much time as you need in order to be as complete as possible. Write a draft on a separate sheet of paper, and take time to revise it before you write the final version here. If you have created a think link rather than a verbal statement, attach it separately.

OPTIONAL ACTIVITY: Visit the Franklin Covey Web site, and complete the interactive personal mission statement for teens:

www.franklincovey.com/cgi-bin/teens/teens-msb/part01/

1.3 Establishing and Tracking Long-Term Goals

The chapter described the importance of goal setting in five different life areas. For each area, name an important long-term goal for your own life. Then, imagine that you will begin working toward each goal. Indicate the steps you will take to achieve your goals on a short-term and long-term basis. Write what you hope to accomplish in the next year, the next six months, the next month, the next week, and the next day.

1.4 Create a Goal-Setting Tree

Write down a long-term goal on an index card or sticky note. Place that at the top of a large piece of newsprint paper. Next, break the goal down into short-term goals. Write the short-term goals on index cards, and place them under the long-term goal. Begin breaking the short-term goals down into smaller, separate tasks, and write these tasks on separate index cards or sticky notes. Place these cards/notes under the appropriate short-term goal. Continue until the tasks are manageable, daily goals that can be placed on a schedule. This will allow you to visualize your journey in a different way.

1.5 Why Go to College?

Why are you considering enrolling in college? Do any of the reasons listed in the chapter fit you? Do you have other reasons all your own? Many people have more than one answer. Write up to five of your reasons.

Take a moment to think about your reasons. Which reasons are most important to you? Why? Prioritize your reasons above by writing 1 next to the most important, 2 next to the second most important, etc.

How do you feel about your reasons? You may be proud of some. On the other hand, you may not feel comfortable with others. Which do you like or dislike, and why?

Endnotes

1 Stephen Covey, *The Seven Habits of Highly Effective People* (New York: Simon & Schuster, 1989), 70–144, 309–318.

2 William B. Werther, Jr., "Workshops Aid in Goal-Setting." In *Personnel Journal* (now *Workforce* magazine), November, 1989.

CONSIDER

CONSIDER

Thinking Critically About College Decisions

Now that you have your SMART goals established, you can begin making many of the decisions that are required before you pack up and move off to college. Thinking critically about these decisions will ensure that your choices will benefit you the most.

If you are overwhelmed with the vast number of decisions you must make over the course of the next year, you are not alone. Thousands of students enter college unsure about what they want to accomplish, why they are at that particular school, and how they will manage a semester (let alone several years) of college. These students often end up taking courses they don't need or want, depleting their financial aid before they have earned a degree, or dropping out. They are often dissatisfied with their college experience and life in general.

However, if you learn the process of critically thinking and evaluating your options, and if you learn to make decisions that follow your personal values and goals, you will find that your college experience, as well as your life, can be satisfying. Learning to take control is a skill you can put to good use throughout your life.

In this chapter, you will explore answers to the following questions:

- What is critical thinking?
- How does my mind work?
- How does critical thinking help me solve problems and make decisions?
- Why plan strategically?

WHAT IS CRITICAL THINKING?

Critical thinking is thinking that goes beyond the basic recall of information. If the word critical sounds negative to you, consider that the dictionary defines its meaning as "indispensable" and "important." Critical thinking is important thinking that involves asking questions. Using critical thinking, you question established ideas, create new ideas, turn information into tools to solve problems and make decisions, and take the long-term view as well as the day-to-day view.

A critical thinker asks as many kinds of questions as possible. The following are examples of possible questions about a given piece of information: Where did it come from? What could explain it? In what ways is it true or false, and what examples could prove or disprove it? How do I feel about it, and why? How is this information similar to or different from what I already know? Is it good or bad? What causes led to it, and what effects does it have? Critical thinkers also try to transform information into something they can use. They ask themselves whether the information can help them solve a problem, make a decision, create something new, or anticipate the future. Such questions help the critical thinker learn, grow, and create.

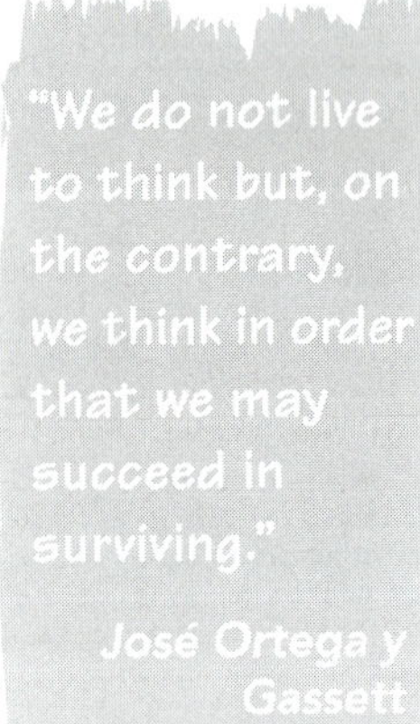

Not thinking critically means not asking questions about information or ideas. A person who does not think critically tends to accept or reject information or ideas without examining them. Table 2.1 compares how a critical thinker and a non–critical thinker might respond to particular situations.

Asking questions (the focus of the table), considering without judgment as many responses as you can, and choosing responses that are as complete and accurate as possible are the ingredients that make up the skill of critical thinking.

Critical Thinking as a Skill

Critical thinking has only recently begun to be taught as such in schools. It used to be assumed that students possessed various levels of thinking ability that would either stay the same or develop naturally in the course of studying particular subjects. Education used to focus primarily on teaching information rather than on how to question and process that information. Now, educators have begun to see critical thinking as a skill that can be taught to students at all different levels of thinking ability. Anyone can develop the ability to think critically.

Learning information is still an important part of education and is, in fact, a crucial component of critical thinking. For instance, part of the skill of

Not thinking critically vs. thinking critically. TABLE 2.1

YOUR ROLE	SITUATION	NONQUESTIONING RESPONSE	QUESTIONING RESPONSE
Student	Instructor is lecturing on the causes of the Vietnam War.	You assume that everying your instructor tells you is true.	You consider what the instructor says, write down questions about issues you want to clarify, initiate discussion with the professor or other classmates.
Parent	Instructor discovers your child lying about something at school.	You're mad at your child and believe the instructor, or you think the instructor is lying.	You ask both instructor and child about what happened, and you compare their answers, evaluating who you think is telling the truth. You discuss the concepts of lying/honesty with your child.
Partner	Your partner feels that he or she no longer has quality time with you.	You think he or she is wrong and defend yourself.	You ask your partner how long he or she has felt this way, ask your partner and yourself why this is happening, and explore how you can improve the situation.
Employee	Your supervisor is angry at your.	You ignore or avoid your supervisor, or you deny responsibility for what the supervisor is angry about.	You are willing to discuss the situation; you ask what you could have done better; you ask what changes you can make in the future.
Neighbor	People different from you move in next door.	You ignore or avoid them; you think their way of living is weird.	You introduce yourself; you offer to help if they need it; you respectfully explore what's different about them.
Citizen	You encounter a homeless person.	You avoid the person and the issue.	You examing whether the community has a responsibility to the homeless, and if you find that it does, you explore how to fulfill that responsibility.
Consumer	You want to buy a car.	You decide on a brand-new car and don't think through how you will handle the payments.	You consider the different effects of buying a new car versus buying a used car; you examine your money situation to see what kind of payment you can handle each month.

critical thinking is comparing new information with what you already know. Your prior knowledge provides a framework within which to ask questions about and evaluate a new piece of information. Without a solid base of knowledge, critical thinking is harder to achieve. For example, thinking critically about the statement "Shakespeare's character King Richard III is like an early version of Adolf Hitler" is impossible without basic knowledge of Shakespeare's play *Richard III* and World War II.

The skill of critical thinking focuses on generating questions about statements and information. To examine potential critical-thinking responses in more depth, explore the different questions that a critical thinker may have about one particular statement.

A Critical-Thinking Response to a Statement

Consider the following statement of opinion: "My obstacles are keeping me from succeeding in school. Other people make it through school because they don't have to deal with the obstacles that I have."

Nonquestioning thinkers may accept an opinion such as this as an absolute truth, believing that their obstacles will hinder their success. As a result, on the road to achieving their goals, they may lose motivation to overcome those obstacles. In contrast, critical thinkers would take the opportunity to examine the opinion through a series of questions. Here are some examples of questions one student might ask (the type of each question is indicated in parentheses):

"What exactly are my obstacles? I define my obstacles as having a heavy class schedule, working evenings and weekends, and being in debt because of my car loan." **(recall)**

"Are there other cases different from mine? I do have one friend who is going through problems worse than mine, and she's getting by. I also know another guy who doesn't have too much to deal with that I can tell, and he's struggling just like I am." **(difference)**

"What is an example of someone who has had success despite having to overcome obstacles? What about Oseola McCarty, the cleaning woman who saved money all her life and raised $150,000 to create a scholarship at the University of Southern Mississippi? She didn't have what anyone would call advantages—money, a college education, membership in the middle or upper class." **(idea to example)**

"What conclusion can I draw from my questions? From thinking about my friend and about Oseola McCarty, I would say that people can successfully overcome their obstacles by working hard and not giving up, focusing on their abilities, and concentrating on their goals." **(example to idea)**

"Who has problems similar to mine? Well, if I consider my obstacles specifically, I might be saying that students who have a heavy class schedule and work many hours will all have trouble in school. That is not necessarily true. People in all kinds of situations may still become successful." **(similarity)**

"Why do I think this? Maybe I am scared of going to college and adjusting to a new environment. Maybe I am afraid to challenge myself or to face the challenges of my college classes. Whatever the cause, the effect is that I feel bad about myself and don't work to the best of my abilities, and that can hurt both me and others who care about me." **(cause and effect)**

"How do I evaluate the effects of this statement? I think it's harmful. When we say that obstacles equal difficulty, we can damage our desire to try to overcome those obstacles. When we say that successful people don't have obstacles, we might overlook the fact that some very successful people have to deal with hidden disadvantages such as learning disabilities or abusive families." **(evaluation)**

Remember these types of questions. When you explore the seven mind actions later in the chapter, refer to these questions to see how they illustrate the different actions your mind performs.

The Advantages of Critical Thinking

Critical thinking has many important advantages. Following are some ways you may benefit from putting energy into critical thinking:

- You will increase your ability to perform thinking processes that help you reach any kind of school, career, or life goal. Critical thinking is a learned skill, just like shooting a basketball or making roses with frosting or using a word processing program on the computer. As with any other skill, the more you use it, the better you become. The more you ask questions, the better you think. The better you think, the more effective you will be when completing schoolwork, managing your personal life, and performing on the job. You will learn more about different critical-thinking processes later in this chapter.

- You can produce knowledge rather than just reproduce it. When you think critically and ask questions, the interaction of new information with what you already know creates new knowledge. When you think critically about lectures or reading materials rather than just learn them for a test, you will retain knowledge that will serve you after you leave school. The usefulness of knowledge comes in how you apply it to new and different situations. It won't mean much for an early childhood education student to quote the stages of child development on an exam unless he or she can make judgments about children's needs when on the job.

- You can be a valuable employee. You certainly won't be a failure in the workplace if you follow directions. However, you will be even more valuable if you think critically and ask strategic questions about how to make improvements, large or small. Questions could range from "Is there a better way to deliver phone messages?" to "How can we increase business to keep from going under?" An employee who shows the initiative to think critically will be more likely to earn responsibility and promotions.

- You can increase your creativity. You cannot be a successful critical thinker without being able to come up with new and different questions to ask, possibilities to explore, and ideas to try. Creativity is essential in producing what is new. Being creative generally improves your outlook, your sense of humor, and your perspective as you cope with problems.

In the next section, you will read about the seven basic actions your mind performs when asking important questions. These actions are the basic blocks you will use to build the critical-thinking processes you will explore later in the chapter.

HOW DOES MY MIND WORK?

Critical thinking depends on a thorough understanding of the workings of the mind. Your mind has some basic moves, or actions, some combination of which it uses each time you think. Sometimes it uses one action by itself, but most often it uses two or more.

Mind Actions and the Thinktrix

You can identify your mind's actions using a system called the Thinktrix, developed by educators Frank Lyman, Arlene Mindus, and Charlene Lopez.[1] They studied how students think and named seven mind actions that are the basic building blocks of thought. These actions are not new to you, although some of their names may be. They represent the ways in which you think all the time.

Through exploring these actions, you can go beyond just thinking and learn how you think. This will help you take charge of your own thinking. The more you know about how your mind works, the more control you will have over thinking processes such as problem solving, decision making, creating, and strategic planning.

Following are explanations of each of the mind actions. Each explanation has the name of the action, words that define it, and examples that explain it. As you read, write your own examples in the blank space provided. Each action is also represented by a picture or icon that helps you visualize and remember it.

Recall: Fact, sequence, description. This is the simplest action. When you recall, you describe facts, objects, or events, or you put them into sequence. Examples:

- Naming the steps of a geometry proof, in order
- Remembering your best friends' phone numbers

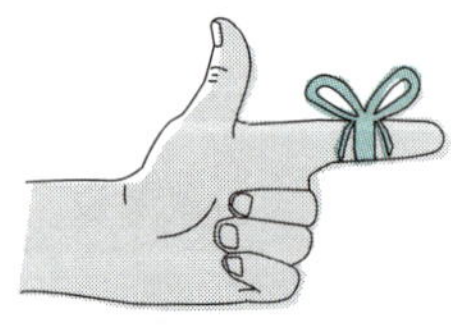

Your example: Recall some facts you know about one college you are considering.

The icon: A string tied around a finger is a familiar image of recall or remembering.

Similarity: Analogy, likeness. This action examines what is similar about one or more things. You might compare situations, ideas, people, stories, events, or objects. Examples:

- Comparing notes with another student to see what facts and ideas you have both considered important
- Analyzing the course catalogs to see how degree plans from two universities are similar

Your example: Tell what is similar about two of the colleges you are considering.

The icon: Two alike objects (in this case, triangles) indicate similarity.

Difference: Distinction, contrast. This action examines what is different about one or more situations, ideas, people, stories, events, or objects, contrasting them with one another. Examples:

- Seeing how two instructors differ in style (one divides the class into small groups and encourages discussion; the other keeps desks in straight lines and lectures for most of the class)
- Contrasting a weekday where you work half a day and go to school half a day with a weekday when you attend class and then have the rest of the day to study

2

Your example: Explain how your educational goals differ from those of your friends.

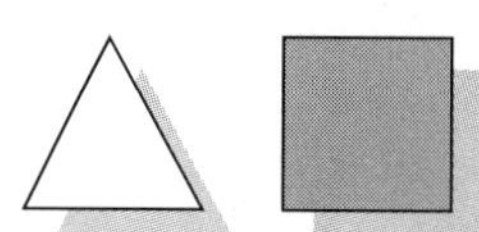

The icon: Two differing objects (in this case, a triangle and a square) indicate difference.

Cause and effect: Reason, consequence, prediction. Using this action, you look at what has caused a fact, situation, or event and/or what effects, or consequences, come from it. In other words, you examine both what led up to something and what will follow because of it. Examples:

- Staying up late at night causes you to oversleep, which has the effect of your being late to class. This causes you to miss some of the material, which has the further effect of your having problems on the test.
- By attending every team practice, you create effects such as being in good physical condition, knowing your teammates' abilities, and being able to think through winning strategies.

Your example: Name what causes you to like your favorite class, and the effects that liking the class has on you.

The icon: The water droplets making ripples indicate causes and their resulting effects.

Example to idea: Generalization, classification, conceptualization. From one or more examples (facts or events), you develop a general idea or ideas. Grouping facts or events into patterns may allow you to make a general statement about several of them at once. Classifying a fact or event helps you build knowledge. This mind action moves from the specific to the general. Examples:

- You are attempting to eat healthy but rarely have time to fix nutritional lunches to bring to school. You know several classmates who completely skip lunch because the cafeteria food is high in fat and calories. Two friends are vegetarians who refuse to buy lunches because all lunches serve some form of meat. From these examples, you derive the idea that your school needs a salad bar.
- You see a movie and decide it is mostly about pride.

Your example: Name examples of activities you enjoy, and from them, come up with an idea of your choice of spring vacation options.

The icon: The "Ex" and arrow pointing to a lightbulb on the right indicate how an example or examples lead to the idea (the lightbulb, lit up).

Idea to example: Categorization, substantiation, proof. In a reverse of the previous action, you take an idea or ideas and think of examples (events or facts) that support or prove that idea. This mind action moves from the general to the specific. Examples:

- When you write a paper, you start with a thesis statement, which communicates the central idea: "Men are favored over women in the modern workplace." Then you gather examples to back up that idea: Men make more money on average than women in the same jobs; there are more men in upper management positions than there are women; women can be denied advancement when they make their families a priority.
- You talk to your high school counselor about attending a college and majoring in accounting, giving examples that support your idea: You have worked in an office after school doing basic bookkeeping tasks, and you have gotten As in all the accounting classes you have taken during high school.

Your example: Name an admirable person. Give three examples of why that person is admirable.

The icon: In a reverse of the previous icon, this one starts with the lightbulb and has an arrow pointing to "Ex." This indicates that you start with the idea, the lit bulb, and then branch into the example or examples that support the idea.

Evaluation: Value, judgment, rating. Here you judge whether something is useful or not useful, important or unimportant, good or bad, or right or wrong by identifying and weighing its positive and negative effects (pros and cons). Be sure to consider the specific situation at hand (a cold drink might be good on the beach in August, not so good in the snowdrifts in January). With the facts you have gathered, you determine the value of something in terms of both predicted effects and your own needs. Cause-and-effect analysis always accompanies evaluation. Examples:

- You decide to try working part-time while still in high school. Your boss has you working twenty to twenty-five hours a week. Often you are responsible for closing and don't get home until after 11:00 P.M. You find that it is difficult to get up early in the morning; you tend to sleep in and then get up too late to attend your first class. From those harmful effects, you evaluate that working while attending high school doesn't work for you. You decide to quit your job and concentrate on passing all your classes this semester.
- Someone offers you a chance to cheat on a test. You evaluate the potential effects if you are caught. You also evaluate the long-term effects on you of not actually learning the material. You decide that it isn't worth your while to participate in the plan to cheat.

Your example: Evaluate your future housing options for college.

The icon: A set of scales out of balance indicates how you weigh positive and negative effects to arrive at an evaluation.

You may want to use a mnemonic device—a memory tool—to remember the seven mind actions. Try recalling them using the word DECRIES—each letter is the first letter of a mind action. You can also make a sentence of words that start with each mind action's first letter. Here's an example: "Really Smart Dogs Cook Eggs In Enchiladas" (the first letter of each word stands for one of the mind actions).

How Mind Actions Build Thinking Processes

The seven mind actions are the fundamental building blocks that your mind uses every day. Note that you will rarely use them one at a time in a step-by-step process, as they are presented here. You will usually combine them, overlap them, and repeat them more than once, using different actions for different situations. For example, when a test question asks you to explain what prejudice is, you might name similar examples that show your idea of what prejudice means.

When you combine mind actions in working toward a specific goal such as making a decision, you are performing a thinking process. The next two main sections will explore some of the most important critical-thinking processes: first, solving problems and making decisions, second, planning strategically amid shifting perspectives.

Each thinking process helps you succeed by directing your critical thinking toward the achievement of your goals. Figure 2.1 shows all of the mind actions and thinking processes together and reminds you that the mind actions form the core of the thinking processes.

The wheel of thinking. FIGURE 2.1

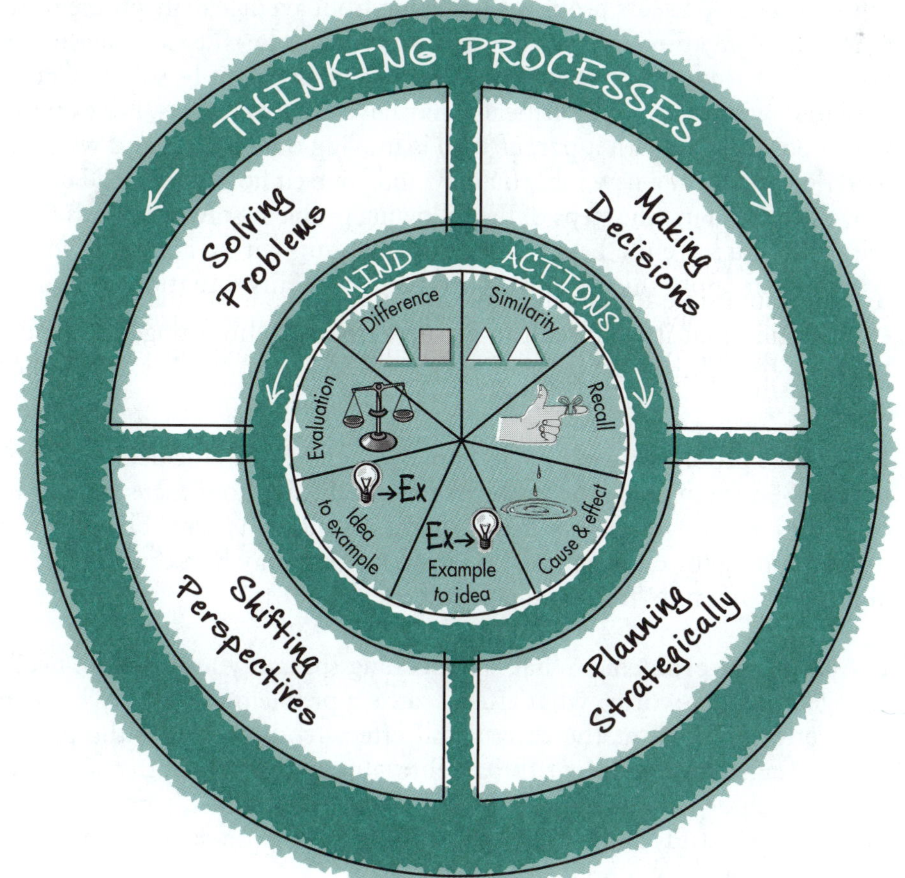

HOW DOES CRITICAL THINKING HELP ME SOLVE PROBLEMS AND MAKE DECISIONS?

Problem solving and decision making are probably the two most crucial and common thinking processes. You are in the process of working through multiple decisions about your future, and each one requires various mind actions. They may overlap somewhat because every problem that needs solving requires you to make a decision. For example, if you are trying to figure out what kind of college to attend, you will be required to make decisions based on the decision you make about what you want to major in. However, not every decision requires that you solve a problem. For example, not many people would say that deciding what to order in a restaurant is a problem. Each process of solving problems and making decisions will be considered separately here. You will notice similarities in the steps involved in each.

Although both of these processes have multiple steps, you will not always have to work your way through each step. As you become more comfortable with solving problems and making decisions, your mind will automatically click through the steps you need whenever you encounter a problem or decision. Also, you will become more adept at evaluating which problems and decisions need serious consideration and which can be taken care of more quickly and simply.

"I have always thought that one man of tolerable abilities may work great changes, and accomplish great affairs among mankind, if he first forms a good plan."

Benjamin Franklin

Problem Solving

Life constantly presents problems—ranging from average daily problems (how to manage study time or learn not to misplace your keys) to life-altering situations (how to manage after your parents are divorced)—to be solved. Choosing a solution without thinking critically may have negative effects. For example, if you choose to live with a parent who is moving out of state and who has to work long hours at a new job, you may find yourself home alone and unhappy. However, if you use the steps of the following problem-solving process to think critically, you have the best chance of coming up with a favorable solution.

You can apply this problem-solving plan to any situation or issue that you want to resolve. Using the following steps will maximize the number of possible solutions you generate and will allow you to explore each one as fully as possible:

1. State the problem clearly. What are the facts? Recall the details of the situation. Be sure to name the problem specifically, without focusing on causes or effects. For example, a student might state this as a problem: "I'm not understanding the class material." However, that may be a cause of the actual problem at hand: "I'm failing my algebra quizzes."

2. Analyze the problem. What is happening that, in your opinion, needs to change? In other words, what effects cause a problem for you? What causes these effects? Look at the causes and effects that surround the problem. Continuing the example of the algebra student, if some effects of failing quizzes include poor grades in the course and disinterest, some causes may include poor study habits, poor test-taking skills, lack of sleep, and not understanding the material.

3. Brainstorm possible solutions. **Brainstorming** will help you think of examples of other similar problems and how you solved them. Consider what is different about this problem, and see if the thoughts you generate might lead you to new possible solutions. It's very important to base your possible solutions on causes rather than effects. Getting to the heart of a problem requires addressing the cause rather than putting a bandage on the effect. If the algebra student were to aim for better assignment grades to offset the low quiz grades, that might raise his GPA but wouldn't address the cause of not understanding the material. Looking at this cause, on the other hand, might lead him to work on study habits or seek help from his instructor, a study group, or a tutor.

Brainstorming
The spontaneous, rapid generation of ideas or solutions, undertaken by a group or an individual, often as part of a problem-solving process.

4. Explore each solution. Why might your solution work? Why not? Might a solution work partially or in a particular situation? Evaluate the pros and cons, or the positive and negative effects, of each idea. Create a chain of causes and effects in your head, as far into the future as you can, to see where you think this solution would lead. The algebra student might consider the effects of improving study habits, getting more sleep, being tutored, or dropping the class.

5. Choose and execute the solution you decide is best. Decide how you will put your solution to work. Then, execute your solution. The algebra student could decide on a combination of improved study habits and tutoring.

6. Evaluate the solution that you acted upon, looking at its effects. What are the positive and negative effects of what you did? In terms of your needs, was it a useful solution or not? Could the solution use any adjustments or changes in order to be more useful? Would you do the same again or not? Evaluating his choice, the algebra student may decide that the effects are good but that his fatigue still causes a problem.

7. Continue to refine the solution. Problem solving is always a process. You may have opportunities to apply the same solution over and over again. Evaluate again and again, making changes that you decide make the solution better. The algebra student may decide to continue to study more regularly but, after a few weeks of tutoring, could opt to trade in the tutoring time for some extra sleep. He may decide to take what he has learned from the tutor so far and apply it to his increased study efforts.

Using this process will enable you to solve personal, educational, and workplace problems in a thoughtful, comprehensive way. Figure 2.2 is a think link that demonstrates a way to visualize the flow of problem solving. Figure 2.3 contains a sample of how one person used this plan to solve a problem.

Decision Making

Although every problem-solving process involves making a decision (when you decide which solution to try), not all decisions involve solving problems. Decisions are choices. Making a choice, or decision, requires thinking critically through all of the possible choices and evaluating which will work best

FIGURE 2.2 Problem-solving plan.

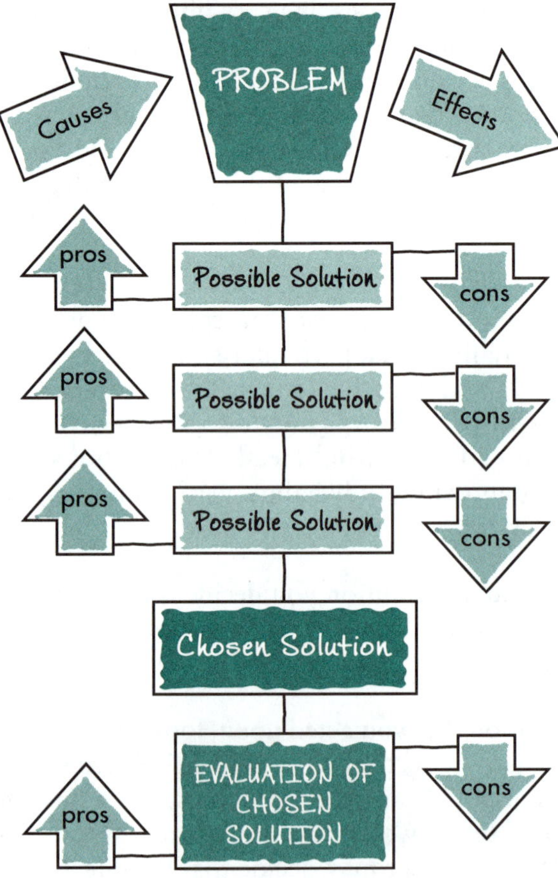

for you and for the situation. Decisions large and small come up daily, hourly, even every few minutes. Do you drop a course? Should you stay in a relationship? Can you work part-time without interfering with school?

Before you begin the decision-making process, evaluate the level of the decision you are making. Do you have to decide what to have for lunch (usually a minor issue) or whether to quit a good part-time job (often a bigger life change)? Some decisions are little, day-to-day considerations that you can take care of quickly on your own. Others require thoughtful evaluation, time, and perhaps the input of others you trust. The following is a list of five steps to take in order to think critically through a decision:

1. Decide on a goal. Why is this decision necessary? In other words, what result do you want from this decision? Considering the effects you want can help you formulate your goal. For example, say a student wants to attend a small private college. Her goal is to become a physical therapist. The school has a good program, but her financial situation makes this school too expensive for her.

2. Establish needs. Recall the needs of everyone (or everything) involved in the decision: The student needs a school with a full physical therapy program;

How one student worked through a problem. FIGURE 2.3

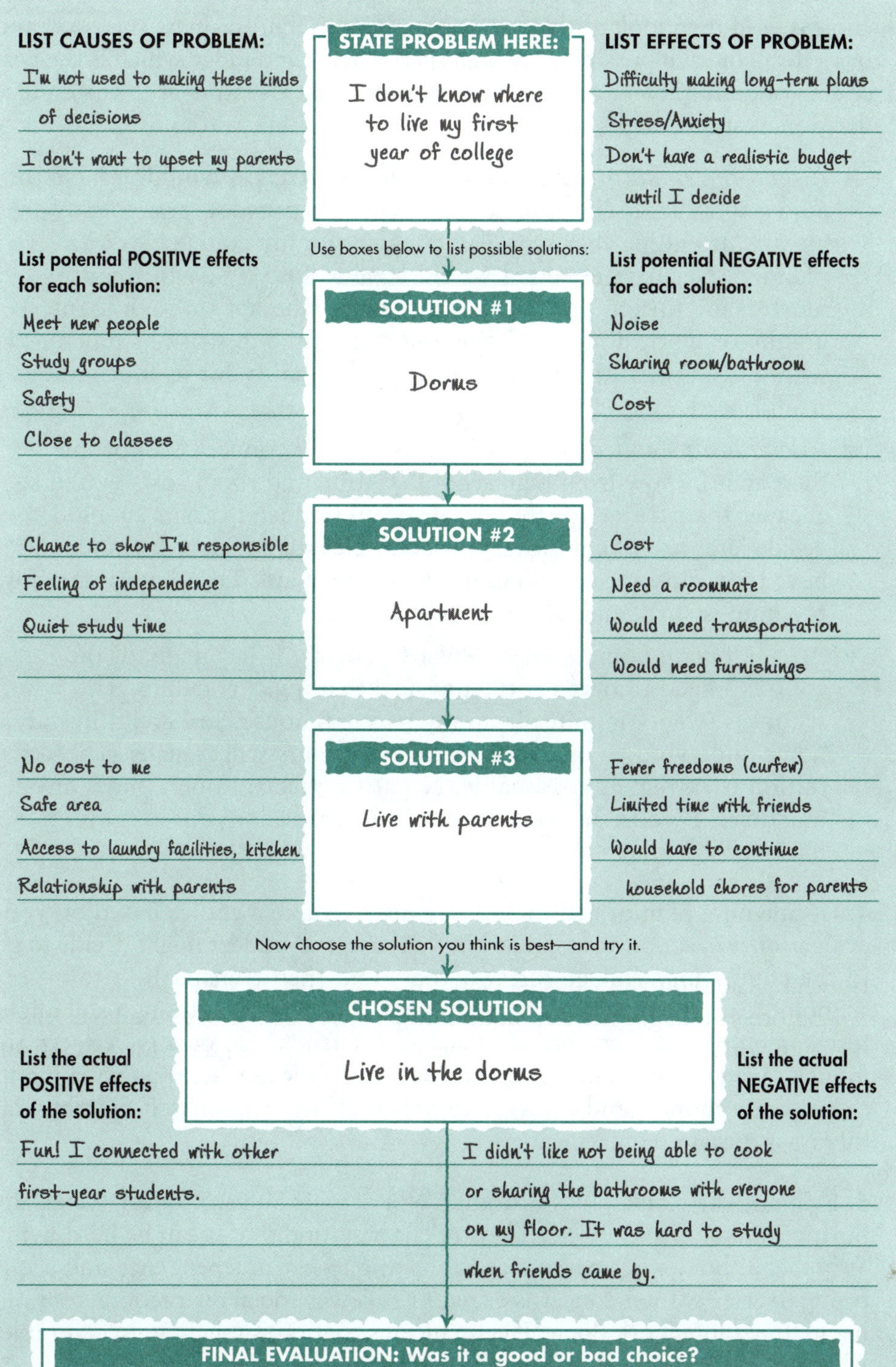

2

she and her parents need to cut costs (her father changed jobs and her family cannot afford the school); she needs to be able to graduate in a timely manner.

3. Name, investigate, and evaluate available options. Brainstorm possible choices, and then look at the facts surrounding each. Evaluate the good and bad effects of each possibility. Weigh these effects and judge which is the best course of action. Here are some possibilities that the student in the college decision example might consider:

- *Attend the small private college.* Positive effects: I wouldn't have to adjust to a large school or to new people because several friends are already attending that school. I could begin my course work as planned. Negative effects: I would have to find a way to finance most of my tuition and costs on my own, whether through loans, grants, or work. I'm not sure I could find time to work as much as I would need to, and I don't think I would qualify for as much aid as I would need.
- *Attend a state college.* Positive effects: I could connect with people there that I know from high school. Tuition and room costs would be cheaper than at the private school. I could transfer credits should I decide later to go to a different school. Negative effects: I would still have to work some or find minimal financial aid. The physical therapy program is small and not very strong.
- *Attend a community college.* Positive effects: It has many of the courses I need to take for the physical therapy curriculum. The school is twenty minutes from my parents' house, so I could live at home and avoid paying housing costs. Credits will transfer. The tuition is extremely reasonable. Negative effects: I don't know anyone there. I would be less independent. The school doesn't offer a bachelor's degree.

4. Decide on a plan of action and pursue it. Make a choice based on your evaluation and act on your choice. In this case, the student might decide to go to the community college for two years and then transfer to a four-year school to earn a bachelor's degree in physical therapy. Although she might lose some independence and contact with friends, the positive effects are money saved, opportunity to spend time on studies rather than working to earn tuition money, and the availability of classes that match the physical therapy program requirements.

5. Evaluate the result. Was it useful? Not useful? Some of both? Weigh the positive and negative effects. The student may find that it can be hard living at home, although her parents are adjusting to her independence and she is trying to respect their concerns as parents. Fewer social distractions result in her getting more work done. The financial situation is much more favorable. All things considered, she evaluates that this decision was a good one.

Making important decisions can take time. Think through your decision thoroughly, considering your own ideas as well as those of others you trust, but don't hesitate to act once you have your plan. You cannot benefit from your decision until you act upon it and follow through.

WHY PLAN STRATEGICALLY?

If you've ever played a game of chess or checkers, participated in a wrestling or martial arts match, or had a drawn-out argument, you have had experience with strategy. In those situations and many others, you continually have to think through and anticipate the moves the other person is about to make. Often you have to think about several possible options that person could put into play, and you consider what you would counter with should any of those options occur. In competitive situations, you try to outguess the other person with your choices. The extent of your strategic skills can determine whether you will win or lose.

Strategy is the plan of action, the method, the how behind any goal you want to achieve. Specifically, strategic planning means having a plan for the future, whether you are looking at the next week, month, year, ten years, or fifty years. It means exploring the future positive and negative effects of the choices you make and actions you take today. You are planning strategically right now just by being in school. You are in the process of making a decision that the requirements of attending college are a legitimate price to pay for the skills, contacts, and opportunities that will help you in the future.

Strategy
A plan of action designed to accomplish a specific goal.

You don't have to compete against someone else in order to be strategic. You can be strategic on your own or even in a cooperative situation. For example, as a student, you are challenging yourself to achieve. You are learning to set goals for the future, analyze what you want in the long term, and prepare for the job market to increase your career options. Being strategic with yourself means challenging yourself as you would challenge a competitor, demanding that you work to achieve your goals with conviction and determination.

What Are the Benefits, or Positive Effects, of Strategic Planning?

Strategy is an essential skill in the workplace. A food company that wants to develop a successful health food product needs to examine the anticipated trends in health consciousness. A lawyer needs to think through every aspect of the client's case, anticipating how to respond to any allegation the opposing side will bring up in court. Strategic planning creates a vision into the future that allows the planner to anticipate all kinds of possibilities and, most importantly, to be prepared for them.

Strategic planning powers your short-term and long-term goal setting. Once you have set goals, you need to plan the steps that will help you achieve those goals over time. For example, a strategic thinker who wants to be in a successful job in five years might make a solid decision about what to major in and what college to attend, take some summer classes, and plan an internship during the senior year of college. Meanwhile, that person will begin to network with potential employers and actively seek opportunities that will increase the opportunity to reach the long-term goal of successful employment. In class, a strategic planner will think critically about the material presented, knowing that information is most useful later on if it is clearly understood.

Strategic planning helps you keep up with technology. As technology develops more and more quickly, jobs become obsolete. It's possible to spend years in school training for a career that will be drying up when you are ready to enter the workforce. When you plan strategically, you can take a broader range of courses or choose a major and career that are expanding. This will make it more likely that your skills will be in demand when you graduate.

Effective critical thinking is essential to strategic planning. If you aim for a certain goal, what steps will move you toward that goal? What positive effects do you anticipate these steps will have? How do you evaluate your past experiences with planning and goal setting? What can you learn from similar or different previous experiences in order to take different steps today? Critical thinking runs like a thread through all of your strategic planning.

Here are some tips for becoming a strategic planner:

- Develop an appropriate plan. What approach will best achieve your goal? What steps toward your goal will you need to take one year, five years, ten years, or twenty years from now?
- Anticipate all possible outcomes of your actions. What are the positive and negative effects that may occur?
- Ask the question "How?" How do you achieve your goals? How do you learn effectively and remember what you learn? How do you develop a productive idea on the job? How do you distinguish yourself at school and at work?
- Use human resources. Talk to people who are where you want to be, whether professionally or personally. What caused them to get there? Ask them what they believe are the important steps to take, degrees to have, training to experience, knowledge to gain.

Strategic planning that stems from conscious goal setting as well as methodical problem solving and decision making will move you closer to your goals in a timely manner. As you set to work mapping your future, periodically refer to the strategies in the chapter to help guide you. Using these strategies will keep you focused toward success.

EXERCISES AND ACTIVITIES

The Seven Mind Actions 2.1

One way to explore the seven mind actions is to apply them to a vocabulary word. Choose a vocabulary word from a course you are taking now. Write your word here.

RECALL. Write the definition. Include two sentences—from the dictionary or from your class materials—that contain the word.

SIMILARITY. What synonyms—words with similar meanings—can you name?

DIFFERENCE. What antonyms—words with opposite meanings—can you name?

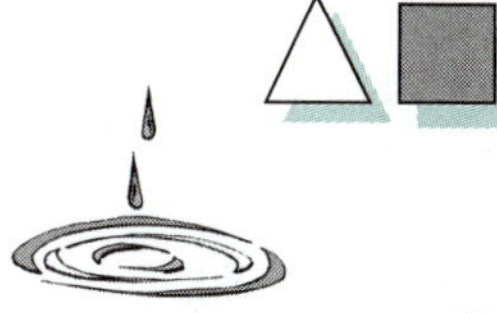

CAUSE AND EFFECT. What effect is caused by using this word—what tone or connotation does it have?

EXAMPLE TO IDEA. From looking at the synonyms and sentences, create a definition in your own words.

IDEA TO EXAMPLE. From the idea or the definition, show an example of the use of the word by placing it in a sentence that you create.

EVALUATION. How well does the word fit in the sentence you have written? Explain.

Making a Decision 2.2

In this series of exercises, you will make a personal decision using the seven mind actions and the decision-making steps described in this chapter. Before you proceed through each of the steps, write an important personal decision you have to make. Choose a decision that you want to act on and will be able to address soon.

Step 1. Name Your Goal

Be specific: What goal, or desired effects, do you seek from this decision? For example, if your decision is a choice between two part-time jobs, the effects you want might be financial security, convenience, experience, or anything else that is a priority to you. It could also be a combination of these effects. Write down the desired effects that together make up your goal. Note priorities by numbering the effects in order of importance.

Step 2. Establish Needs

Who and what will be affected by your decision? If you are deciding how to finance your college education you must take into consideration your family's financial needs as well as your own when exploring options. List the people/things/situations that may be affected by your decision, and indicate how your decision will affect them.

Step 3. Check Out Your Options

Look at all the options you can imagine. Consider options even if they seem impossible or unlikely—you can evaluate them later. Some decisions only have two options (e.g., when you go to college, to move into an apartment or not, to get a roommate or not); others have a wider selection of choices. On your paper, first list the possible options for your own personal decision. Then evaluate the good and bad effects of each.

Have you or someone else ever made a decision similar to the one you are about to make? What can you learn from that decision that may help you?

Step 4. Make Your Decision and Pursue It to the Goal

Taking your entire analysis into account, decide what to do. Write your decision.

Next is perhaps the most important step: *Act on your decision.*

Step 5. Evaluate the Result

After you have acted on your decision, evaluate how everything turned out. Did you achieve the effects you wanted to achieve? What were the effects on you? On others? On the situation? To what extent were they positive, negative, or some of both? List four effects. Name each effect, write Positive or Negative, and explain that evaluation.

Final evaluation: Write one statement in reaction to the decision you made. Indicate whether you feel the decision was useful or not useful, and why. Indicate any adjustments that could have made the effects of your decision more positive.

Endnote

1 Frank T. Lyman, Jr., "Think-Pair-Share, Thinktrix, and Weird Facts: An Interactive System for Cooperative Thinking." In *Enhancing Thinking Through Cooperative Learning,* ed. Neil Davidson and Toni Worsham (New York: Teachers College Press, 1992), 169–181.

INVESTIGATE

INVESTIGATE

3

Exploring Careers and Majors

Although many students come to college knowing what they want to study, many do not. That's completely normal. High school and college are perfect times to begin exploring your different interests. In the process, you may discover talents and strengths you never realized you had.

Some of your explorations may take you down paths that don't resonate with your personality and interests, but each experience will help you to clarify who you really are and what you want to do with your life. In fact, because our society is changing so rapidly, career exploration should become a lifelong pursuit, and the techniques of career exploration you learn in this chapter can be used over and over again.

In this chapter, you will explore answers to the following questions:

- When should I choose a career?
- What should I know about the future job market?
- How do I explore careers?
- How are careers and majors related?
- What type of college should I attend?

WHEN SHOULD I CHOOSE A CAREER?

Nearing high school graduation, you may be feeling pressured into making a decision about what career to pursue. However, at this point in your life, you may not be prepared to make that kind of a decision. You may not even be aware of all the career possibilities that exist, especially since so many new careers are created yearly! Unfortunately, in the United States, our society does pressure young adults to make career decisions that can have a lasting effect. What happens if you choose the wrong career? Or what if you want to change your mind after a few years of study?

It may help ease the pressure to know that the U.S. Department of Labor has estimated that "young people of high school age should expect to have an average of fourteen jobs throughout their lifetimes, in possibly six to eight different career fields."[1] What a change from a generation ago when individuals often began working right out of high school and worked for the same company until they retired!

What does this prediction about the job market mean for you? On the positive side, it probably means you will never be locked into one particular job. You will have the opportunity to try new skills, learn new information, and experience new adventures throughout your life. On the other hand, it will mean you will have to have or develop the skill of being flexible with your career. You will need to learn how to transition from one job to another with minimum time, training, and possibly even education.

If you are unsure about what career you wish to pursue, relax. Career counselors will tell you that it is okay not to be locked into a career decision at your high school graduation. In fact, they will most likely urge you to take some time to self-reflect, explore your options, take exploratory classes, and make an informed decision when you feel comfortable doing so.

But, taking the pressure off doesn't mean you should stop thinking about your options now. This is the perfect time to begin exploring what the future holds in terms of employment and to begin reflecting on your personal values, interests, and skills.

WHAT SHOULD I KNOW ABOUT THE FUTURE JOB MARKET?

Although no one can predict the future perfectly, we can look at trends and get a general picture of what the job market will be like a few years down the road. The Bureau of Labor Statistics offers the following information for the years 1996–2006[2]:

- Industry employment growth is projected to be highly concentrated in service-producing industries, with business, health, and education services accounting for 70 percent of the growth:
    - Health care services will increase 30 percent and account for 3.1 million new jobs.

- Educational services are projected to increase by 1.3 million teaching jobs.
- Computer and data processing services will add more than 1.3 million jobs.

- The labor force will become increasingly diverse:
  - The labor force growth of Hispanics, Asians, and other races will be faster than that for blacks and white non-Hispanics, stemming primarily from immigration.
  - Women's share of the labor force is expected to increase from 46 percent to 47 percent.
- Jobs will be available for job seekers from every educational and training background:
  - Almost two-thirds of the projected growth will be in occupations that require less than a college degree. However, these positions generally offer the lowest pay and benefits.
  - Jobs requiring the least education and training—those that can be learned on the job—will provide two of every three openings due to growth and replacement needs.
- Job growth varies widely by educational and training requirements:
  - Occupations that require a bachelor's degree are projected to grow the fastest. All of the twenty occupations with the highest earnings require at least a bachelor's degree. Engineering and health occupations dominate the list (see Figure 3.1).
  - Education is essential for getting a high-paying job. However, many occupations—for example, registered nurses, supervisors of blue-collar workers, electrical and electronic technicians, automotive mechanics, and carpenters—do not require a four-year college degree, yet they offer higher-than-average earnings.

These projected trends give only a brief profile of what the future job market will be like. One thing does seem certain: The more technological skills you learn, the better your chances of landing a high-paying job. Figure 3.1 illustrates predictions about occupations that will have fast growth, high pay, and largest numerical growth; Figure 3.2 illustrates job growth rates based on education and training.

Don't despair if you are passionate about a career that isn't listed in the top twenty-five future jobs. If you *are* passionate about a specific career, you should pursue that goal! Although you might find that you have to work harder at finding a job, that you have to work more than one job until you can pursue your dream full-time, or that you don't earn as much money as others, you will be happy in your profession. Greater job satisfaction will reflect positively in other areas of your life.

Additionally, if you diversify your skills, have a well-rounded background, and continue to be a lifelong learner, you will find that you have more opportunities, and that it is easier to move into a wide variety of jobs.

FIGURE 3.1 Occupations with largest numerical growth.

Occupations with fast growth, high pay, and low unemployment that have the largest numerical growth, projected 1996–2006.

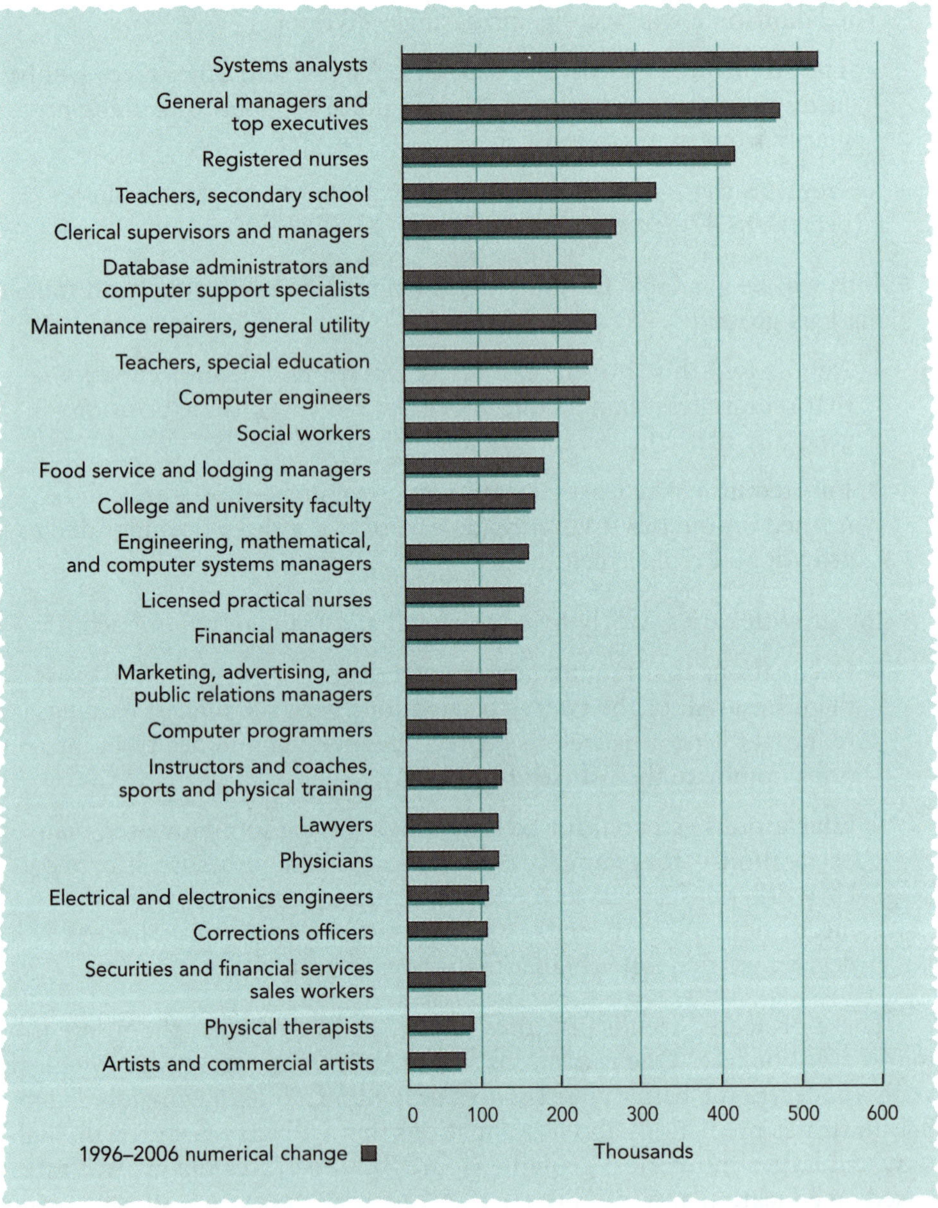

- Systems analysts
- General managers and top executives
- Registered nurses
- Teachers, secondary school
- Clerical supervisors and managers
- Database administrators and computer support specialists
- Maintenance repairers, general utility
- Teachers, special education
- Computer engineers
- Social workers
- Food service and lodging managers
- College and university faculty
- Engineering, mathematical, and computer systems managers
- Licensed practical nurses
- Financial managers
- Marketing, advertising, and public relations managers
- Computer programmers
- Instructors and coaches, sports and physical training
- Lawyers
- Physicians
- Electrical and electronics engineers
- Corrections officers
- Securities and financial services sales workers
- Physical therapists
- Artists and commercial artists

0 100 200 300 400 500 600

Thousands

1996–2006 numerical change ■

Source: *1998–1999 Occupational Outlook Handbook.* Washington, DC: U.S. Department of Labor.

HOW DO I EXPLORE CAREERS?

As a high school student, this may be the first time you've seriously explored your career options; however, career exploring and planning may occur many times over the course of your life. As society advances, new opportunities are created. If you continue to explore careers and keep up-to-date with your skills and education, you will be ready when something new and challenging comes along.

Growth rates by education.

Growth rates by most significant source of education and training, projected 1996–2006.

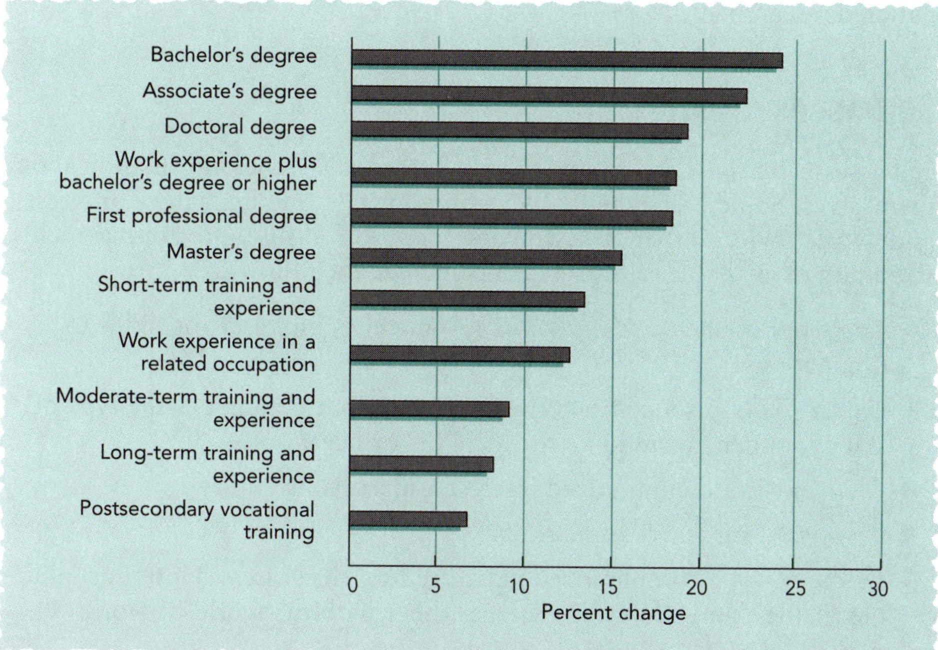

Source: *1998–1999 Occupational Outlook Handbook.* Washington, DC: U.S. Department of Labor.

One of the most important things to remember right now is that you do have time. In fact, the first two years of college are generally spent fulfilling general education requirements such as math, writing, communications, arts, science, and social sciences. These freshman- and sophomore-level classes give students an opportunity to strengthen critical-thinking skills and develop a solid background in the basics, which will help them succeed in upper-level classes and give them the opportunity to explore various fields of study.

As you begin college, you will have several opportunities to meet with an academic advisor to discuss career goals and academic planning. If, at an advising session, you are given a degree plan that you are uncertain about, don't worry. You are not locked into pursuing that degree. Advisors often give you this degree plan, or a number of degree plans, to get you to think about your college and career goals. These degree plans list all of the courses a student in that field of study is required to complete in order to earn a certificate or degree. Once you have declared a major, these degree plans act as a contract between you and the college. If you are following an established degree plan and the school revises and changes the degree plan, you will probably not be required to take additional classes, only the classes listed on your original degree plan. Often, the first two years of any degree plan are similar because every student is completing the general education requirements.

Although you are free to change degree plans whenever you want, you should be advised that the best time to change majors is before you take a great number of upper-level courses. Often, the credits for one degree don't match those for another, so you end up taking extra semesters to complete your degree. If you are using financial aid to pay your college expenses, the money may run out, and you will have to pay for the extra classes on your own.

Choosing a degree plan, or career, can be a difficult task; however, you can begin the exploration process right now so that the decision becomes less complicated. Two steps you should consider are self-assessments and occupational research.

Self-Assessments

Self-assessment tools are created to help individuals gather information about themselves. Some assessments focus on your personality, values, interests, and work styles; other assessments focus on skills and competencies. Examples of available self-assessments include the following[3]:

- *Campbell Interest and Skill Survey*—matches interests and skills to occupations.
- *Career Skills*—is a computerized program that determines the type of skills a student would like to use in his or her work.
- *Choices*—is a computerized interest and skills inventory.
- *Compass*—measures basic skills.
- *Discover*—is a computerized guidance tool that assesses interests, abilities, values, and skills and matches those with the world of work.
- *FOCUS II*—identifies interests, skills, and values and relates them to occupations.
- *Myers-Briggs Type Indicator*—determines personality and matches to work styles.
- *Self-Directed Search*—matches interests and abilities to career fields.
- *SIGI Plus*—is a computerized guidance tool that looks at interests, personality, values, and occupational options.
- *Strong Interest Inventory*—matches interests to occupations.

Another valuable self-assessment tool is the Prentice Hall Self-Assessment Library on CD-ROM.

These assessments will ask you questions about yourself, your interests, your values, and your skills. Your high school counseling office, the U.S. Department of Labor, or the career counseling office at any college can provide you with a number of self-assessments.

You can also choose to create your own self-assessments. For example, you could:

- Look at your personal mission statement to reflect on your personal values.
- Create a table that lists all your accomplishments. For each accomplishment, list the skills you used in that activity.
- Journal write about the things that interest you, your aspirations, and people you admire.
- Compare jobs you have had or would like to have. Create a list of pros and cons associated with each job.

- Create a list of your personal traits and characteristics, and think about what jobs might match your traits.
- Join activity clubs and organizations to see what interests you. Evaluate what you like or dislike about your experience.
- Take a class that focuses on career exploration. Your high school, local adult basic education, community education center, or local college may offer these courses.
- Complete a personality-style inventory, and compare the results to various job requirements.

The process of self-assessment is ongoing because we grow and change with each new experience. What you value now may not be as important in a few years, and you need to take these changes into account when you are considering careers.

After completing one or more self-assessments, you should have a self-profile that you can match to various jobs. If you work through specific programs in career centers, your self-assessment tool may give you a computer-generated list of suggested careers that would suit your profile. Think of this as a starting place for more specific types of research about careers and occupations.

Occupational Research

In addition to self-assessments, you should also plan to spend time researching occupations. When we think of research, we often think of reading through stacks of books; however, research can take many forms. Your research may include looking through books and occupational guides, but you may also want to interview someone in the job you are considering, shadow someone in the job, take on part-time work in that field, volunteer, or try a cooperative education experience.

> "To accomplish great things, we must not only act, but also dream, not only plan, but also believe."
> Anotole France

There are specific points of information you should gather during your research. For example, you will want to find out answers to the following questions:

- How much knowledge or training would I need to be hired for this job?
- What kinds of other skills would I need to learn (communication skills or team-building skills, for example)?
- What kinds of responsibilities would I have in this job?
- What are the working conditions of this job?
- What opportunities for advancement would I have in this job?
- What is the salary range for this job?
- What is the future outlook for this job?
- What are similar jobs that I might consider?

The place to begin researching jobs is your high school counseling office or library. Most likely, it will have specific occupational guides that you can browse through, as well as numerous books, videos, and computer programs that offer comparable information. As you complete your research, take notes and compare the various jobs you research, but don't feel pressured into making a decision immediately. Leave your options open as you work through the exploration stage of career planning.

Interviewing people performing the job you are interested in is a great way to obtain more information and makes the information from the books real for you. You can ask questions that might not have been answered in your previous research, as well as questions that pertain to specific geographic jobs. For example, the research might show you that there are a large number of available jobs nationwide, but in an interview, you might discover that there are a very limited number of job openings in your city or state. Interviews can also open the door for you to shadow someone performing the job.

Job shadowing

Working side by side with someone as a learning experience.

Job shadowing is an opportunity for you to work side by side with someone who is hired in that profession. Depending on the type of job, you might find yourself observing others, or you may even get the chance to help. Benefits of job shadowing include discovering both the good and bad aspects of a particular job, learning about the job environment, and making contacts with others in the profession. Job shadowing may also open the door for you to be hired in part-time employment.

One of the best ways to find out if a particular type of work is suited to you is to do it. There are several ways: part-time or full-time paid work and volunteer work. At this point in your education, you may find that you are too busy with academic and extracurricular activities to take on a job, and that's okay—you are learning valuable skills in school, organizations, and sports. But, when you evaluate your time, you may discover that you could work one or two days after school or one day of the weekend. You don't want to overload yourself, but even a minimal amount of work experience will certainly help you decide whether you like certain jobs and teach you work-related skills such as communication, leadership, team building, and problem solving. You may even get referrals for future jobs.

Cooperative education provides you with the opportunity to have paid employment in positions that complement your academic program. Although most co-op positions are obtained at the college level, more and more high schools are taking the initiative and offering high school students co-op positions. In these cases, students generally take classes for half a day and work for half a day.

Christopher Pratt, director of career services and pre-professional advising for the Massachusetts Institute of Technology and the Atlantic Cooperative Education Training and Resource Center, tells us there are seven reasons to co-op: "Co-op students do better in school; are more likely to graduate; are ahead in preparation for their field; are viewed as better candidates in the job interviews by employers; receive more job offers; earn higher starting salaries; and are more likely to get the job with the employer they want after graduation."[4]

If you have the opportunity to complete a co-op program, make the most of it by developing learning objectives in consultation with your supervisor, monitoring your progress, and making changes to your goals and strategies, if necessary.

However you choose to research occupations and careers, take time to reflect on your experiences. What do you like or dislike about the job? Do you like the work environment and the pace of the job? Do you have the skills, or can you easily learn the skills, to make you successful in the job? Does the job challenge or bore you? Would you be happy in the job for more than a few years? Does the career offer multiple yet similar job opportunities that you could take advantage of?

As you analyze these questions, consider whether you want to pursue this type of job. If not, consider yourself lucky that you discovered that answer now, not five or ten years down the road. If you do think this is a career worth pursuing, it may be time to think about what to major in at college.

HOW ARE CAREERS AND MAJORS RELATED?

Once you have decided on the type of career you want to pursue, you should research the type of skills and education you will need. The required education will indicate what type of degree you will need and what you will need to major in. A major is a group of classes that are required to earn a degree in a specific area. Some careers, such as heating and air-conditioning technology or business and office technology, may only require a one-year certificate or a two-year degree. These types of careers take very specialized training and education, so you will need to declare your major early in your academic career so that you can get into the classes you will need.

Some careers don't require specific majors in entry-level positions. For example, a person entering a career in marketing might major in marketing, accounting, communications, or public relations. As you research your career options, find out what type of major or degree is required so you can plan accordingly. Additionally, this information should help you decide what type of college to attend.

WHAT TYPE OF COLLEGE SHOULD I ATTEND?

There are many types of colleges from which to choose: two-year community or four-year public or private colleges and universities. People choose colleges based on different reasons, but one of the major deciding factors should be the type of degree or major you want to pursue for your chosen career.

Community Colleges

Community colleges are two-year colleges that offer associate degrees and may even offer various certificate programs that take only a year to complete. Additionally, community colleges usually have reduced tuition, compared to four-year colleges. For example, the tuition at a community college may be $35 per credit hour, but the tuition at a four-year college across town may be $100 per credit hour. Community colleges are becoming more and more popular because of their lower-than-average costs and their reputations as good learning and teaching institutions.

Many students who plan to earn a bachelor's degree also begin their academic careers at a community college. Typically, bachelor's degrees take four years to complete, but if students plan carefully, they may be able to earn the first two years at a community college and then transfer to a four-year college. This type of planning can save students a tremendous amount of money and is a good way to transition into college and explore options.

Other students may discover that although they have earned a general education diploma (GED) or high school diploma, their skills may not be strong enough for entry into four-year general education classes. Community colleges generally offer a wide variety of development courses in reading, writing, and math. Students can enroll in these courses in order to prepare themselves for the general education requirements. When community colleges and four-year colleges are located within the same city, students sometimes find themselves attending both at the same time, taking preparatory classes at the community college and other classes at the four-year college.

Four-Year Colleges and Universities

Four-year colleges or universities offer many different degree programs, primarily in the bachelor's degree areas; some four-year colleges also offer associate degrees and provide various graduate-level—master's and doctoral—programs as well. Although university tuition is much more expensive than community college tuition, there are benefits as well. Universities usually offer more resources in financial aid as well as student support such as advising and library services. Universities typically offer more in the way of student life, such as student groups, campus social events, collegiate and intramural sports, and cultural events such as concerts, guest speakers, and plays.

Universities are usually divided into smaller colleges such as the College of Business, the College of Agriculture, or the College of Arts and Sciences. These colleges oversee and administer the degree programs specific to their academic area; students who major in business administration would be advised in and then take most of their classes in the College of Business. Some of their classes, such as the general education core and various electives, would be taken in other colleges.

Many different types and sizes of four-year colleges exist; there are some that enroll many thousands of students and those that serve a few hundred. There are colleges that specialize in one or two areas and some that offer a broad range of degree options. Many public colleges offer what is called open admission: Students may be admitted if they meet certain minimum standards. Private colleges have stricter admission standards and, usually, higher tuition, but they may offer certain specialized benefits or learning environments; for example, many private colleges are affiliated with certain religious denominations.

As you explore career possibilities, remember that you do not have to commit to a single idea, college, or career right this minute. Keep focused on what you value and your long-term goals. These two things should guide you as make decisions about your life and as you begin shopping for a college that fits those needs.

EXERCISES AND ACTIVITIES

Assessing My Strengths, Talents, Abilities, and Interests 3.1

Complete three of the self-assessment activities listed in the chapter. When you have finished, evaluate what you learned about yourself. Are there any trends that you notice? Match your assessment with possible career choices, and begin researching those options.

OPTIONAL EXERCISE: Use the Prentice Hall Self-Assessment Library CD-ROM to explore your skills, abilities, and interests.

Interviewing for Information 3.2

Interview a person working in the field you are interested in pursuing. Formulate your questions so that you get as much relevant information as possible. When you have finished your interview, evaluate the information you acquired. What were the positive or negative aspects about this job? Do you still think you might want to pursue this type of career?

Degree Plan Evaluation

Contact the student advising center of some of the colleges you are considering, and request degree plans for the majors you are interested in. Evaluate those degree plans, and create a list of classes you should take in high school to prepare you for that major or degree.

Majors Exploration

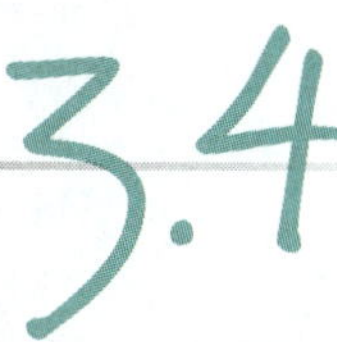

Visit the Prentice Hall Supersite at **http://www.prenhall.com/success/majors/index.htm** and explore possible majors. Use this information as a springboard into other types of occupational research.

Endnotes

1 "Basic Skills of Career Exploration." www.dmdc.osd.mil/asvab/CareerExplorationProgram/main.html (September 4, 1999).

2 "Tomorrow's Jobs." *1998–1999 Occupational Outlook Handbook.* stats.bls.gov:80/oco/oco2003.htm (September 4, 1999).

3 "How to Start Thinking About a Career, Launch a Job Search, and Have a Future Beyond College." *Planning Job Choices: 1999,* 42 (1999): 6.

4 Christopher Pratt, "Cooperative Education and Internships: A Window into the World After College." *Planning Job Choices: 1999,* 42 (1999): 53.

D E T E R M I N E

DETERMINE

4

Choosing a College that Fits Your Needs

One of the most challenging decisions a prospective college student has to make is which academic institution to attend. Students today have more options to choose from than ever before. Since our society has moved from an industrial society to a society of technology and information, the workforce's demand for trained employees has prompted the need for employees to have a college education. The result has been an increase in the number of colleges willing to train students to meet the needs of the workforce.

How does this affect you? Although it may seem like you have a mind-boggling amount of options to choose from and decisions to make, having so many choices is actually to your benefit. As you've probably noticed from all the college brochures and videos, most colleges want you to attend their institution. One key to having a satisfying college experience is to make sure you are at a college that fits your needs, matches your value system, and moves you in the direction of meeting your long-term goals.

Even though it may seem tempting simply to attend either a local state college that many of your friends are attending or your parents' alma mater, it is essential that you take control of your decision and choose a college based on your personal goals.

How do you find a college that fits your needs? You have to shop around! Think of college as a major purchase. Most people don't buy the first car off the lot without checking out other cars and dealerships. The same thing should be true as you shop for colleges. Because your primary reason for going to college is to get a degree, the most important factor to consider is a college's academic reputation. However, other factors can also impact your decision. You should consider student services and activities (including athletics), library and technological resources, tuition expenses, and housing expenses and options.

In this chapter, you will explore answers to the following questions:

- Why should I shop around when selecting a college?
- What types of academic information should I consider?
- What types of student services and activities should a college provide?
- What should I learn about tuition?
- What are some housing options and expenses?
- What resources can help me make an informed decision?

WHY SHOULD I SHOP AROUND WHEN SELECTING A COLLEGE?

Because choosing a college is like making a major purchase, it is crucial to shop around in order to find the best possible option for you. All of these factors—academics, student activities and organizations, tuition, and housing expenses—are factors that deserve serious consideration when you are making the decision about where to attend college.

As you look through college brochures and catalogs or listen to their videos, keep in mind that there are several factors on which to base your decision; it is important to decide where your personal priorities lie. For instance, if money is not a problem for you, then tuition and housing costs probably are not strong factors in your decision. But for most students, money is a factor, and these types of expenses may have more weight than other areas that will be discussed.

First of all, always keep your long-term goals in mind as you make your decision. Secondly, use your critical-thinking skills to evaluate the colleges based on academics; student services, organizations, and activities; tuition; and housing.

WHAT TYPES OF ACADEMIC INFORMATION SHOULD I CONSIDER?

Degree Programs

Clearly, a student should select a college that is going to serve his or her needs in the best way possible. One of the first options you should research is whether a college you are considering offers majors in your area of interest. Not all colleges offer all types of majors—that would be redundant and impossible! Often, state institutions offer programs that emphasize majors that are different from those of other institutions in the state. For example, one state college may emphasize medicine, another may emphasize engineering, while yet another college may emphasize education. If you are not yet sure what you want to major in, you should select a college that most closely emphasizes your interests. Remember that two- and four-year degree programs require general education requirements that students take in their first two years. During this time period, you will have the opportunity to more thoroughly explore majors, and if it is necessary, you can transfer to a different college that offers the degree you desire.

Transfer Options

Students sometimes choose to attend a two-year community college the first two years. This is a viable option to consider when selecting a college. If you do plan to attend a two-year college and then transfer to a four-year college, you need to be very careful in selecting classes. Two-year colleges generally have articulation agreements, which means that the four-year colleges will automatically accept credit for specific classes taken at two-year schools. It is your responsibility to get in writing a list of classes that can be easily transferred. Once you have this list, don't stray from it! Otherwise, you will find yourself repeating classes, incurring additional costs, and delaying your graduation.

Faculty Reputation and Research

Depending on your major area of study and the type of degree you are pursuing, faculty reputation and research may be a concern for you. If you are interested in knowing about the faculty, the best way to get some answers is to visit them! Make appointments to see them and talk to them about their work—this will let them know that you are a prospective student who is serious about your future!

If, however, you are not able to visit faculty in person, you should visit through the college home pages on the Web. Many instructors post not only office information but also course syllabi, schedules, and specific assignments. These sites will give you at least a general idea of what to expect if you should enroll in one of those classes.

Accreditation

All legitimate colleges go through an accreditation process. They are evaluated by independent accrediting agencies that periodically review the school's curriculum, standards, and results.

Additionally, programs within the college should be accredited in their specific discipline. Specific professional programs should be accredited by the appropriate accrediting agency in their field. For example, a nursing program should be accredited by the National League for Nursing, or an EMT/paramedic program should be accredited by the Joint Review Committee on Educational Programs for the EMT-paramedic. If you can't find information about the accreditation, you should ask. Making sure you are in a top-notch program is ultimately your responsibility.

If the college you are considering is not accredited by a regional accrediting association or the programs within the college are not accredited, you should probably consider a different college.

Scholarship, Work-Study, Internship, Co-Op, and Job Placement Opportunities

Because college is a costly venture, you should find out what types of financial aid and employment opportunities each college provides and how to apply.

Scholarships can help a great deal when you are faced with college expenses because they provide financial assistance that does not need to be repaid. Some scholarships are based on financial need; other scholarships may be based on special talents or academic performance. Generally, you must submit financial aid applications in order to apply for any type of financial aid. Once you've received the required paperwork, apply for as many scholarships as possible. Even small scholarships can help pay for books and supplies that you will need.

"Knowing is not enough; we must apply. Willing is not enough; we must do."
Goethe

Work-study jobs are another way to help with college expenses; in addition, you may get lucky enough to land a job that will look good on your resume when you graduate. These jobs are located on the college campus, and the supervisors often attempt to work around students' class schedules. Work-study jobs may range from five to twenty hours a week. When you visit a campus, check out the campus job board and see if you might qualify for any. These jobs go quickly, so the sooner you apply, the better chance you will have.

Intern and co-op positions give students an opportunity to work in the career field in which they are studying. Although internships and co-ops are all different, the point of these programs is to give students an opportunity to apply the knowledge they have gained in the classroom while gaining on-the-job experience. Intern and co-op programs place students in jobs in the community and in work with professionals in their field. Many positions pay very well, and students are sometimes offered full-time positions when they graduate. A college with a healthy intern or co-op program is worth serious consideration—it demonstrates that the college has a realistic view of employer expectations and employment opportunities.

Finally, in terms of academics, you should research the school's job placement rate. Find out how many graduates are employed after graduation. If there is a high job placement rate, it indicates that the school is well respected from an employer's point of view.

As you are critically thinking about your choices based on academic qualities, make sure the colleges provide the information presented in Table 4.1.

Academic considerations.

TABLE 4.1

Does each college provide the following academic considerations?

- The academic degree that I want.
- Faculty who focus on students and who are current in their professional field.
- Appropriate accreditations.
- Scholarships, work-study opportunities, internships, co-ops, and job placement programs.

WHAT TYPE OF STUDENT SERVICES AND ACTIVITIES SHOULD A COLLEGE PROVIDE?

The quality and quantity of student services and activities can tell a prospective student a great deal about the college. Student services that cater to special needs and populations should be obvious, and a variety of activities that build a sense of community should appear on a campus calendar.

Student Services

Some **student services** are basic to every college campus—advising centers, financial aid offices, tutorial services, career development and placement services, counseling centers, and libraries. These are key offices to visit when making your college selection. Personnel in these services who are clearly student-friendly, professional, knowledgeable, and up-to-date reflect a campus with the same qualities.

Student services
Services provided by a college to help students be successful, including academic advising, career counseling, tutoring, financial aid advising, and many others.

A thorough investigation also includes researching auxiliary student services and organizations. Auxiliary services might include a student health center, eating facilities, technological laboratories, and cultural programs. Certainly if you are interested in pursuing a degree in computer science, you want to attend a college that is committed to technology and provides up-to-date computer labs and software. Likewise, if you want to major in Latin American studies, you should attend a campus that clearly reflects a commitment to cultural studies and programs.

Student Organizations

Student organizations can be a key factor in your college experience. Becoming involved in student organizations is one way to connect with your campus and make you feel a part of that community. Because there is such a wide variety of student organizations, you should be able to find one that interests you. For example, you may want to join a fashion merchandising club, a water technology club, a drama club, or an intramural sports team. Your involvement in student organizations demonstrates your commitment to an idea and allows you the opportunity to work with other individuals with similar beliefs and values. Furthermore, student organizations provide key opportunities to strengthen your leadership skills, and future employers may be very interested to hear about your involvement.

Student Athletics

For some students, athletic programs have no weight in making a decision about which college to attend, but for others, athletics are a major factor. If you are basing your decision on athletics, you should consider the following:

Graduation rates for athletes. How many athletes in the college have graduated during the last five years? If there is a low graduation rate, that school probably isn't the best choice. After all, what good is going to college if you don't get a college degree? A few athletes have the ability to turn professional, but most don't; in the long run, you would be better off to be on a team that has a sound reputation and record for stressing academics and graduation.

Program completion time. How long does it take most athletes to complete a program of study? Although it is common for athletes to take longer than regular full-time students to complete a program of study, that time shouldn't be excessive. If student athletes are taking longer than five or six years, it's a good bet that academics aren't stressed until after eligibility has expired.

Scholarships. What kinds of scholarships are available for student athletes? If you truly have a talent to offer the university and are capable of successfully completing your academic commitment, you should expect that university to offer some scholarship or financial assistance. An athletic recruiter should be able to answer your questions about financial assistance clearly—get it in writing *before* you sign any letters of intent.

NCAA probationary status. What is the National Collegiate Athletic Association probationary status of the team you are considering? If you find yourself being recruited by a team that is serving probation because of violations, find out the cause of the probation. If this team has broken rules under the current coaching staff, you should probably not spend a great deal of time considering this team. If, however, the violations occurred under a different staff, you may be okay. Perhaps they are in the process of rebuilding a program, and you could be instrumental to the team.

Athletic status. What will your athletic status be as a first-year student? Will you be redshirted? Or will you be an active team member? These are questions that may affect financial aid and scholarships, so you should have a clear idea about your status before you commit to a team.

Special student athletic services. What types of special services are given to student athletes? Because of demanding practice and traveling schedules, student athletes can have difficulties keeping up with academic demands. It is important to find out if your team offers services such as tutorial programs designed especially for athletes.

Practice and traveling schedules. How long will you be expected to practice each day, and how extensive are the travel schedules? If you discover that your schedule will be difficult, you may need to make a tough decision about whether you want to participate in collegiate athletics. After all, your goal is to get a solid education and degree, so decide where you are willing to concentrate your efforts.

Student Activities

In addition to student organizations, a college should sponsor student activities that provide a sense of campus community. Perhaps there are homecoming celebrations, holiday events, special concerts, or movie nights. Although these activities might not be the deciding factor in your college decision, they do play an important role in campus living and provide not only entertainment but a sense of campus pride.

WHAT SHOULD I LEARN ABOUT TUITION?

Cost is a fundamental concern when selecting a college or university. One of the two major expenses is **tuition,** the price your courses will cost you per credit hour. A few things to keep in mind when looking at tuition costs include full or part-time tuition and in-state or out-of-state tuition.

Tuition
The cost of your courses per credit hour.

Tuition expenses can range from very low to tens of thousands of dollars and are usually presented in two different ways: the total cost of tuition for full-time students and the cost of tuition per credit hour for other students.

Full-Time Status

College classes are presented by credit hours, depending on the amount of time spent in class, and a full-time student takes at least twelve credit hours per semester—usually four classes. If a full-time student chooses to enroll in more than twelve credit hours, there is usually no extra charge for the extra credit hours unless a student takes eighteen or more credit hours—not something a first-year student should even consider. Additionally, health and activity fees are also included in the cost of full-time tuition. Depending on the college, these fees may cover the cost of such privileges as using the campus health center; attending sports, music, and theater events; and using special campus facilities such as computer labs. Meal plans may be available, but the price of the meal plan is not included in the tuition fee, nor is the price of textbooks.

Part-Time Status

If you will not be able to attend classes on a full-time basis, you will pay your tuition fees by a credit-hour rate. For example, if you take two classes that are each three credit hours, you will pay for six hours of tuition. If the tuition costs $150 per credit hour, you will pay $900 to take those two classes. Health and activity fees are not included in the credit-hour rate but may be available at an additional fee.

In-State Tuition

In-state tuition simply means that you will be attending a college located in the state in which you reside. These fees are substantially lower than out-of-state tuition. For example, one college lists its in-state tuition as $1,200, but out-of-state tuition at the same college is $4,100.

Out-of-State Tuition

Out-of-state tuition is sometimes referred to as nonresident tuition. Residential requirements vary from state to state. For example, colleges located on state borders may have special agreements to accept students from neighboring cities in the adjoining state, so if you are considering attending a college in a different state, be sure to find out these requirements. Out-of-state tuition may be four or five times more expensive than in-state tuition, so sometimes it may be financially beneficial to attend an out-of-state college on a part-time basis until residency has been established. Additionally, some colleges offer special scholarships to be used specifically to cover this added expense.

Does Tuition Reflect the Quality of Education I Will Receive?

Although a $30,000-a-year school is likely to be viewed as more prestigious than a $10,000-a-year school, does a degree from a prestigious school really give a graduate a substantial competitive edge in the job market? Opinions differ, but you should remember that employers are looking for well-educated applicants, not applicants with expensive degrees. What you choose to do with your educational opportunities is more important than the price tag of your tuition. Many very successful individuals have started their college careers at less expensive community colleges and then transferred to a four-year college.

WHAT ARE SOME HOUSING OPTIONS AND EXPENSES?

Tuition may seem like it should be the largest expense you pay, but the biggest expense is often housing if you choose a campus in a different city or state than where you live. As an in-coming first-year college student, your housing options may be limited. Some campuses require that first-year students live on campus and in specific dorms; however, other campuses may not have these types of requirements. Before deciding where you want to live, you should consider the pros and cons of all options.

Residential Halls

Although some students groan at the thought of living in the residential halls, or dorms, the reality is that residential life offers many advantages for first-year students. Living in the residential halls allows you to meet and make new friends, participate in residential team-building activities, and live in a protected and safe environment that is close to all your collegiate activities.

Residential regulations vary from campus to campus, but generally there are several options from which students can choose. For example, you may be able to live in a room that you share with only one roommate, or you may choose to live in a suite with several other students. Additionally, residential halls are sometimes reserved solely for certain groups of students, such as athletes, women, or honor students.

If you are a person who needs a great deal of privacy and solitude, the residential halls may not be the best choice for you. But for many students, the residential halls give them the opportunity to make lasting friendships, to connect with student tutors and mentors, and to sharpen people skills.

ΣΠΩ

Fraternity and Sorority Houses

Although images from the movie *Animal House* may come to mind when you think of living in a fraternity or sorority house, fraternities and sororities do offer a viable housing option that rarely reflects the movie. This option, however, may not be available until your sophomore year or even later, depending on the fraternity or sorority. And there are varying eligibility requirements for joining fraternities and sororities that should be taken into consideration before considering this an option.

The living arrangements in fraternity and sorority houses are often similar to residential halls in which you have two or more roommates. In some instances, the members reside in "sleeping porches," very large rooms that house all members.

4

The cost of living in a fraternity or sorority is sometimes comparable to living in residential halls but can sometimes be much more expensive, depending on the organization. Be sure to thoroughly research this expense if you do decide to pledge.

Apartments

Most students are excited at the prospect of living in an apartment for the first time, and with careful shopping and planning, apartment living may be the least expensive housing option. Apartments can offer privacy and independence that residential halls and fraternity or sorority houses can't, but that privacy and independence can come with a higher price tag than is expected. For example, you may have to pay utilities, security deposits, and transportation costs to get to and from school, and you may have to pay extra for a furnished apartment.

Apartments can cost you in other ways, as well. For example, if you are a first-year student in a new city, you may not know any other students. Living alone in an apartment does not offer you the opportunity that you would have in a residential hall to easily meet other students. And, by living in an apartment, you may have to sacrifice some of the safety that comes with living on campus. However, if you share the apartment with one or more roommates, these financial and social expenses may seem reasonable, and apartment living might be your best choice.

Parents' or Relatives' Homes

The very least expensive housing option is to continue to live at home with your parents or to live with a relative. Often, you can live free and have the added bonus of having meals with your family and access to conveniences such as laundry facilities! Even if you are required to pay rent, it is usually much less than you would have to pay elsewhere.

One disadvantage of living at home or with relatives may be the lack of the degree of independence that other students have. For example, if your friends are living in the residential halls and have freedom to stay out as long as they want, you may be tempted to do the same. Sometimes parents aren't willing to give college students that much independence.

If you choose to live at home or with a relative, it is imperative that you sit down and discuss expectations before problems arise. Parents may be more willing to compromise and bend their rules if you discuss this with them prior to following through with your plans.

Because housing is one of the greatest expenses you'll encounter as a college student, it is important to research the options carefully for each college you consider. Your choice needs to be livable—both financially and socially. For example, if you are a person who is extremely shy and it is difficult for you to meet others, living in an apartment could further isolate you and make your college experience unbearable. Weigh your options carefully and be fair to yourself.

Table 4.2 asks you to consider some important questions before making a decision about housing.

As you gather information about the colleges you are considering, keep track of it using Table 4.3 at the end of the chapter. Make multiple copies of the form so you can compare the colleges you're considering.

TABLE 4.2 Where will I live?

OPTIONS	QUESTIONS TO CONSIDER
Residential Hall	Will I live with someone I know or someone I haven't met before? How will I manage distractions from other residents?
Fraternity or Sorority House	Will I be able to manage my schoolwork, time, money, and fraternity activities effectively? Will I be able to say no to fun activities when I have tests to study for and papers to write?
Apartment	Will I share the apartment with a roommate? How will I meet friends and get involved in campus life?
Home	Will I have the same freedoms that I would have if I were living elsewhere? Will I be expected to pay rent or have other household responsibilities?

WHAT RESOURCES CAN HELP ME MAKE AN INFORMED DECISION?

Collecting the information you need to make an informed decision may seem like an overwhelming task; however, most of the information can be found in a few key places.

Much of the initial information can be found in college catalogs, which list detailed information about degree programs, classes, tuition and housing expenses, and some student services. This type of information can also usually be found online by clicking on the colleges' home pages.

Many online services exist for the sole purpose of helping you compare institutions. These services are free and provide a wealth of information. If you use an online service, carefully check its sources of information, data collection methods, and sponsors. The following sites can help you make informed decisions:

Petersons—**www.petersons.com**

College Board—**www.cbweb1.collegeboared.org/cohome.htm**

CollegeNET—**www.collegenet.com**

U.S. News & World Report's College Ranking—
www.usnews.com/usnews/edu/college/corank.htm

The Princeton Review Online—**www.review.com/college/templates/temp2.cfm?topic=rank&body=rank/index.cfm&Link=rank.cfm&special=College.cfm**

Money Online: Value Rankings for Colleges—
www.pathfinder.com/money/colleges98/article/rankindx.html

University of Illinois Library Collection—
www.library.uiuc.edu/edx/rankings.htm

"We can avoid making choices by doing nothing, but even that is a decision."
Gary Collins

Once you have narrowed your choices, it is imperative that you visit the college campuses and meet with individuals who can answer specific questions for you. Before arriving, you should make appointments to see representatives in the offices such as financial aid, student advising, housing, and your major area of study. These individuals can help answer your questions and provide you with key information that will help you make the most informed choice.

TABLE 4.3 College comparison worksheet.

	COLLEGE 1	COLLEGE 2	COLLEGE 3
Location ■ distance from home			
Size ■ enrollment ■ physical size of campus			
Environment ■ type of school (2 yr., 4 yr.) ■ school setting (urban, rural) ■ location & size of nearest city ■ co-ed, male, female ■ religious affiliation			
Admission Requirements ■ deadline ■ tests required ■ average test scores, GPA, rank ■ notification			
Academics ■ your major offered ■ special requirements ■ accreditation ■ student-faculty ratio ■ typical class size			

(continued)

College comparison worksheet (continued).

TABLE 4.3

	COLLEGE 1	COLLEGE 2	COLLEGE 3
College Expenses ■ tuition, room & board ■ estimated total budget ■ application fee, deposits			
Financial Aid ■ deadline ■ required forms ■ percent receiving aid ■ scholarships			
Housing ■ residence hall requirement ■ food plan			
Facilities ■ academic ■ recreational ■ other			
Activities ■ clubs, organizations ■ fraternity/sorority ■ athletics, intramurals ■ other			
Campus Visits ■ when ■ social opportunities			

EXERCISES AND ACTIVITIES

4.1 Comparing and Contrasting Colleges

Collect catalogs from colleges you are considering. Research information from these catalogs, and compare and contrast the facts about academic programs, tuition, housing, etc. Use copies of the checklist (see Table 4.2) to keep your information organized.

When you have collected the information, write a proposal to your parents, explaining to them why you have chosen a particular college.

4.2 Interviewing Recruiters

Make appointments with recruiters from colleges you are considering. Write down questions you would like them to answer, and interview them.

When you have completed the interview, write a letter to yourself that explains the positive and negative aspects of attending each college.

4.3 Visit a College Campus

If possible, visit two or three colleges you are considering. Spend a day exploring the campus and learning more about the campus environment. If possible, spend that time with a college student you know, and attend classes and activities with him or her.

4.4 College Life and Housing

Assemble a panel of college students to speak to you about their colleges and their college experiences. Talk to them about what it is like to live in the residential halls, fraternity or sorority houses, or apartments.

4.5 Shopping for Facts

If you are considering living in an apartment, get newspapers from that town and "apartment shop." Research hidden costs such as transportation and utilities. Put together a week's menu and "shop" for the food.

When you have completed your "shopping," make a list of your expenses to see if living in an apartment is a viable option for you.

Part II Keys to College Admission

CONSTRUCT

5

Building an Academic Portfolio

Throughout this text, you will be asked to do a great deal of self-reflection and writing. Already you've considered your value system and personal mission statement, as well as long-term and short-term goals. You've learned how to explore careers and think critically about decisions and to problem-solve. Additionally, you will practice skills such as note taking and time management that will help you when you enter the college environment.

How do you compile all of this information so that it reflects your accomplishments, skills, and goals? One successful means of presenting yourself to others is through the portfolio.

In this chapter, you will explore answers to the following questions:

- What are portfolios?
- How are portfolios used?
- Why should I build a portfolio?
- When should I begin to build a portfolio?
- What information can be found in a portfolio?
- How do I build a portfolio?
- How do I keep my portfolio strong?

WHAT ARE PORTFOLIOS?

A portfolio is a collection of work that represents an individual's abilities and accomplishments. Historically, artists have always compiled portfolios of their work. They used these portfolios to show prospective employers and buyers samples of their work. The portfolio allowed them to showcase their skills.

Today, the concept of using portfolios has spread into all facets of the workforce. For example, it is highly possible that your elementary, middle school, and high school teachers have been creating a portfolio of your work since you started school. In this folder are samples of the work you did at school. Teachers often use the portfolios to assess the strengths of students. It also gives teachers an idea of what students know and what concepts need to be emphasized.

Portfolios are used not only in education but in many career fields as well. Currently, the military uses the concept of portfolios for individuals in the armed forces who want to be promoted to a higher rank. These individuals put together packets that demonstrate what they have accomplished, the education and skills they have, and any other evidence that shows they should be promoted to a higher rank.

HOW ARE PORTFOLIOS USED?

Portfolio

A collection of information and documents that reflect an individual's abilities and accomplishments.

Portfolios are used in many different ways. Sometimes they are used during the hiring process. Prospective teachers often assemble portfolios that contain sample lesson plans, sample assignments, and other evidence that they are good teachers. This collection of work is something that prospective employers can use to decide whether the teacher is a good match for their school.

Portfolios may also be used to determine promotions and salary increases. The newspapers are filled with cartoons showing individuals begging their bosses for raises and bosses routinely rejecting their pleas. Today when individuals negotiate pay raises and promotions, it is common for them to use a portfolio to help them win their case. It is more difficult to say no when strong evidence is presented using a professional method such as a portfolio.

Self-assessment is another way in which portfolios are used. When individuals collect documents that reflect their accomplishments, strengths and weaknesses, and goals, it becomes easier to evaluate progress toward goals. Using portfolios as personal self-assessment is one way to evaluate and refine your personal mission statement and your long- and short-term goals.

WHY SHOULD I BUILD A PORTFOLIO?

You may find several uses for your personal portfolio, including applying for admission to colleges, specific college programs, scholarships, and jobs. For instance, when applying for admission to any college, you will need to submit specific documents. (More information about these documents can be found in Chapter 6.) As you are collecting these documents, you are creating a portfolio that profiles you and your achievements. For example, you may want to include documents that reflect your leadership in a specific project or an award you received. Portfolios give you the opportunity to present yourself in the best way possible and will certainly give you an edge over those who present themselves in an unorganized way.

Often, students don't recognize their own strengths as individuals until they create a portfolio. Most of us are busy individuals, participating in many different kinds of activities and excelling in different areas of our lives. When we think about the whole picture and how much we accomplish overall, we are often surprised.

Going through the process of assembling a portfolio gives individuals the chance to reflect and to see what patterns exist in their lives. As you put together your portfolio, you may suddenly realize that you have more leadership experience than you ever thought. Perhaps you've taught swimming lessons, directed a puppet theater for the summer reading program, and served as the athletic trainer for the basketball team. These are all legitimate examples of leadership.

WHEN SHOULD I BEGIN TO BUILD A PORTFOLIO?

You may be saying to yourself, "Why do I need to worry about creating a portfolio now? I'm not going to apply for jobs until *after* graduation." That *may* be true, but because there are so many uses for portfolios, the time to start building one is now. Remember that a portfolio is a collection of your work and accomplishments, so you need to collect those documents as you complete them.

The wrong time to start building a portfolio is the night before a deadline! Building a professional-looking portfolio takes time. As you are applying for admission and scholarships, you will need to collect specific documents, such as transcripts and test scores. Because you are collecting these documents from various sources, the process may take weeks.

Start today by requesting any documents you may need and by reflecting on what you have done that will demonstrate the kind of person you are. Find the evidence that will prove your abilities.

WHAT INFORMATION CAN BE FOUND IN A PORTFOLIO?

There isn't just one set of guidelines for assembling and using portfolios. As you go through life, you will need to customize your portfolio, depending on its purpose. Not only might the contents change, but also the form of the portfolio.

All portfolios are different, depending on their purpose. For example, a person who is using a portfolio for promotion purposes has much different information in the portfolio than a person who is applying to serve in the Peace Corps.

Furthermore, your portfolio may take on different forms. Someone who is applying for a job as a Webmaster for a large corporation would probably choose to create a digital or electronic portfolio; a person applying for an accounting position in the same corporation may have a more traditional portfolio.

Additionally, you may choose to customize your portfolio based on the way you use it. As you are searching through scholarships to apply for, you will want to note not only their required documents but also the values they desire. When you are aware of your audience and their expectations, it is easy to tailor your work to their desires. This is not to say that you will lie in your portfolio, but rather that you will emphasize some skills or accomplishments over others, and you may even choose not to include some material.

For example, if you are applying for a scholarship that is based on academic merit, you would naturally want to showcase achievements in that area. However, if you are applying for a scholarship that is based on service to the community, you would want to discuss how you've volunteered at the local Boys and Girls Club, led a campaign to introduce a recycling program in your neighborhood, and participated in a fund-raiser for juvenile diabetes. See Table 5.1 for questions you should consider when customizing your portfolio.

You may feel like you have little to put in your portfolio at this time. After all, perhaps you've never had a paying job or won any state competitions. Don't let that stop you. If you've been actively participating in academics and in your school activities, you probably have plenty to include. Following are suggested ideas that you might include in your personal portfolio, depending on its purpose and audience.

5

A copy of your personal mission statement and long- and short-term goals. An admissions counselor, scholarship committee, or prospective employer would already know a great deal about you and what you value by reading your personal mission statement. Having stated goals and a plan of action for reaching those goals impresses others. It shows you have reflected on what is

TABLE 5.1 Customize your portfolio.

PURPOSE	*Why are you creating this portfolio?* Application to a particular school or academic program? Is it for a job? A scholarship? An award?
AUDIENCE	*Who will be reading this portfolio?* Will it be supervisors? A scholarship committee made up of faculty, staff, and students? Peers? Faculty? Community members?
FORMAT	*How should I present this information?* Should it be presented in a notebook? A folder? Electronically?
REQUIRED DOCUMENTS	*Have I included all the required documents that have been requested?* Can I, or should I, include other relevant documents? Will I be penalized for including additional documents? Have the documents been revised or updated?
OTHER INFORMATION	*What information should be contained in a cover letter that explains the portfolio?* Do I need multiple copies of the portfolio? Who are the individuals I can contact if I have questions?

important to you (your values) and made decisions about how to live your life according to those values.

A copy of your resume. Even though you may not have had many paying jobs, you should include those you have held, as well as any volunteering you have done and projects you have worked on for organizations you belong to. For example, if you were the recording secretary for an organization for two years, you should list that. It demonstrates your commitment to the organization, as well as your leadership potential, organizational skills, and communication skills. Your resume doesn't need to be elaborate, but it does need to be clearly written so that others can glean information about you from it.

Copies of transcripts, your diploma, and any certifications you have earned. This information would be appropriate when applying for admission and scholarships; however, it might not always be appropriate. Use your best judgment when including this information.

Copies of any awards you have received. If needed, include an explanation about the award. Often the award itself is explanation enough and is telling evidence of your personal character and abilities.

Copies of recommendation letters. If you have excelled in particular classes or have done exceptional work for an individual, consider asking for a letter of recommendation. These letters could be rather general letters that describe the relationship you have with the individual (this person's student for two years, for instance), a description of the work you have accomplished, your skills, and general information about your character. If you need specific information for a specific purpose, don't hesitate to tell the person so the letter can be most effective.

Copies of names of references and their contact information. References are people who will vouch for you and your skills. They may be contacted and asked specific questions about your abilities. Make sure the contact information—phone numbers, mailing address, e-mail address—is kept current. Also make sure you get permission to use them as references. It is an uncomfortable situation for someone to be called and asked to give a reference when that person is not expecting it. The opposite is also true: If the person named as reference is expecting to be called, he or she can be prepared to discuss your achievements and give a strong, positive profile of you.

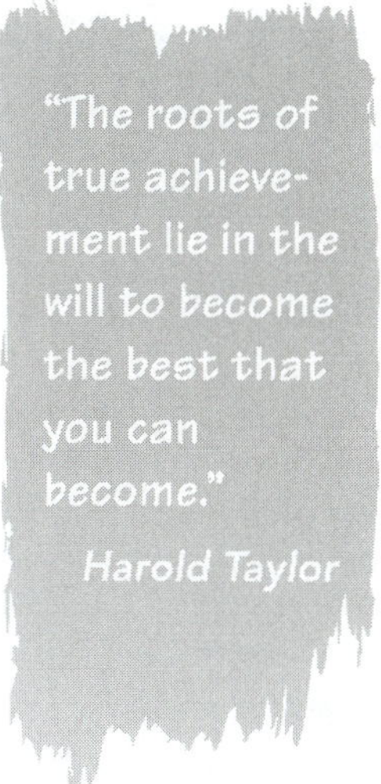

Copies of your work samples. Admissions counselors, scholarship committees, and prospective employers often want specific examples of work you have completed. Outstanding writing samples are very helpful, so you might consider including a copy of an essay or article you wrote. Group projects are also appropriate if you include a description of your participation and leadership in the project. Also consider including a piece of work that demonstrates your level of critical and creative thinking. Perhaps you designed an advertising campaign for your yearbook. Include copies of some of the work you created.

Any other requested information or materials that will showcase your skills. For example, if you are planning on majoring in early childhood education, you would want to find a way to demonstrate your skills in working with children. You might write a summary of your experiences that describes how you've learned to effectively manage caring for children of various ages, how you've learned to solve problems, and how you completed a study on children's nutrition and snacks.

Your portfolio might look a little different every time you use it. Always keep in mind the purpose of the portfolio when you are selecting items to include in it.

Following are some other suggestions to keep in mind:

- If you are sending your portfolio to someone, include a cover letter that explains why you are sending the portfolio and a brief description that highlights the contents.
- Put your materials in a logical order. If you are responding to a specific scholarship application that asks for specific materials, put the materials in the order in which they are listed on the application.
- Include the appropriate information and the appropriate amount of information. You want the person reviewing your portfolio to get a clear and complete profile of you, but you don't want to overwhelm that person. If you make him or her wade through excessive information, that person may not bother to look at any of it. Be complete, but don't go over the limit.
- If you include a great deal of information, find a way to make it accessible. For example, you might include tabs or staple sections separately.
- Keep your materials current. As you grow as a student, the work you produce will reflect that growth. Your thinking, writing, and leadership skills will strengthen, and you want the work in your portfolio to reflect that growth. Exchange your old examples for new ones.
- Keep your references current. For example, as you eventually move through college and get ready to enter the job market, you will replace the letter from your high school forensics coach with a letter from a college instructor. Likewise, when you work for different employers, always ask them for letters of recommendation or for permission to use them as references.
- Make sure your portfolio looks neat and professional. With today's easy access to computers, there really isn't a reason to include handwritten cover letters, resumes, or other information. This will be especially true when you approach graduation from college and will use your portfolio in the job market.
- Have your portfolio critiqued by an individual who can give you good advice. The process of assembling a portfolio is much the same as writing an essay. You should go through the process of having the portfolio critiqued and revised in order to present a high-quality profile of yourself.

HOW DO I BUILD A PORTFOLIO?

Do you remember that old adage, "Rome wasn't built in a day"? The same is true for effective portfolios. You may have tried to write a paper the night before it was due or study for a test an hour before taking it. What was the

result? Was the paper the best it could be? Did you get every question on the test correct? Probably not. Building a strong portfolio also takes time, and like a paper you write for your English class, it probably will need to be revised—possibly more than once!

As you begin to create your portfolio, think of its purpose in general terms. This should be a collection that you can pick and choose from when you are assembling portfolios for specific reasons.

You should probably consider investing in a small file that you can keep your materials in. Most of the time, you will want to send copies of documents, instead of originals, so you should have separate folders for each document. Make sure you mark the original in some way so that you won't accidentally send it away. Keep a few copies of the original ready in case you need to assemble multiple copies of the portfolio at one time. This is especially helpful if you are going to apply for admission to several different colleges or for multiple scholarships.

You should spend some time brainstorming your accomplishments and activities. At this point in the process, don't edit yourself or leave anything out. It's best to gather as much information as possible before you decide what is important and what isn't. When law enforcement agencies are investigating a crime, they are required to collect every type of information possible before they actually present the case. Think of your portfolio as evidence that proves your abilities; you also should collect as much information as possible before presenting your case.

Following are suggestions for collecting information:

- Fine-tune your personal mission statement, and keep a copy in your files. Even if you don't use it in all the portfolios you send out, having it and using it will keep you focused on your goals.
- Get copies of transcripts and test scores from your school.
- Begin drafting your resume. If you don't know how to write a resume, check with your counselor or English teacher, who should have a packet of information for you. Or, you can purchase one of the many how-to books at your local bookstore.
- Consider carefully whom you might ask to write a letter of recommendation for you. Choose three or four individuals, and talk to them personally about what your goals are and why you would like them to write a letter for you. You might consider asking teachers, employers, club or activity sponsors, or adults who know you well. When they have written their letters, be sure to thank them.
- Make a list of all the awards you have earned, and make copies of the certificates that accompany the awards. Don't forget to include community service recognition as well as school activities.
- Sort through completed school assignments that demonstrate your academic abilities. Choose ones that emphasize your thinking and writing abilities.

Once you have a collection of materials ready, create a sample portfolio that you can have critiqued. Teachers or counselors who know you well would be good people to ask because they may remember something you've done but haven't included. After they have looked at it and given you sug-

gestions for improvement, begin to revise. If your reviewers are willing to look at it again, let them. When you are happy with the materials, file them away until you need to assemble a portfolio for a specific purpose.

After completing your portfolio, you should be able to reflect on your accomplishments with a sense of pride and confidence. You will discover how valuable your work as a student, volunteer, participant, and leader has been. By creating a portfolio, you showcase not only your accomplishments as an individual but also your qualities and character. This should give you the motivation and self-confidence to move ahead with your life.

HOW DO I KEEP MY PORTFOLIO STRONG?

Even when you get accepted into college or get the scholarship or a job you want, your work with your portfolio won't be over. You should consider your portfolio a living document that needs to steadily grow as you do. As you improve your skills and your thinking and as you participate in new experiences, you should document these accomplishments and add this evidence to your growing portfolio file. And as your older material becomes out-of-date and irrelevant, remove it from your files.

One way to keep your portfolio growing is to create and then take advantage of opportunities that you excel in. For example, you could find a campus organization to participate in and volunteer to be an officer, or you could join a community service organization. And, of course, you could take a co-op, internship, or job that will prepare you for the career you want after college graduation.

Creating a portfolio now will keep you organized and ready for any opportunity that may come your way.

EXERCISES AND ACTIVITIES

My Personal Mission Statement 5.1

Revisit the personal mission statement you wrote in Chapter 1. Refine this mission statement, and use it for your portfolio.

College Admission Portfolios 5.2

Assemble a portfolio that you can use in applying for college admission. Include the materials suggested in this chapter, as well as any other materials you think will be helpful.

When you have assembled your portfolio, write a cover letter that explains your purpose for presenting the portfolio.

Scholarship Portfolios

Research scholarship contests, and use your portfolio to apply. Reassemble your portfolio so that it contains the required materials, and rewrite your cover letter so that it is appropriate for your purpose.

Career Portfolios

Research part-time job possibilities, and use your portfolio to apply. Reassemble your portfolio and rewrite your cover letter so that both are appropriate for the job.

B E G I N

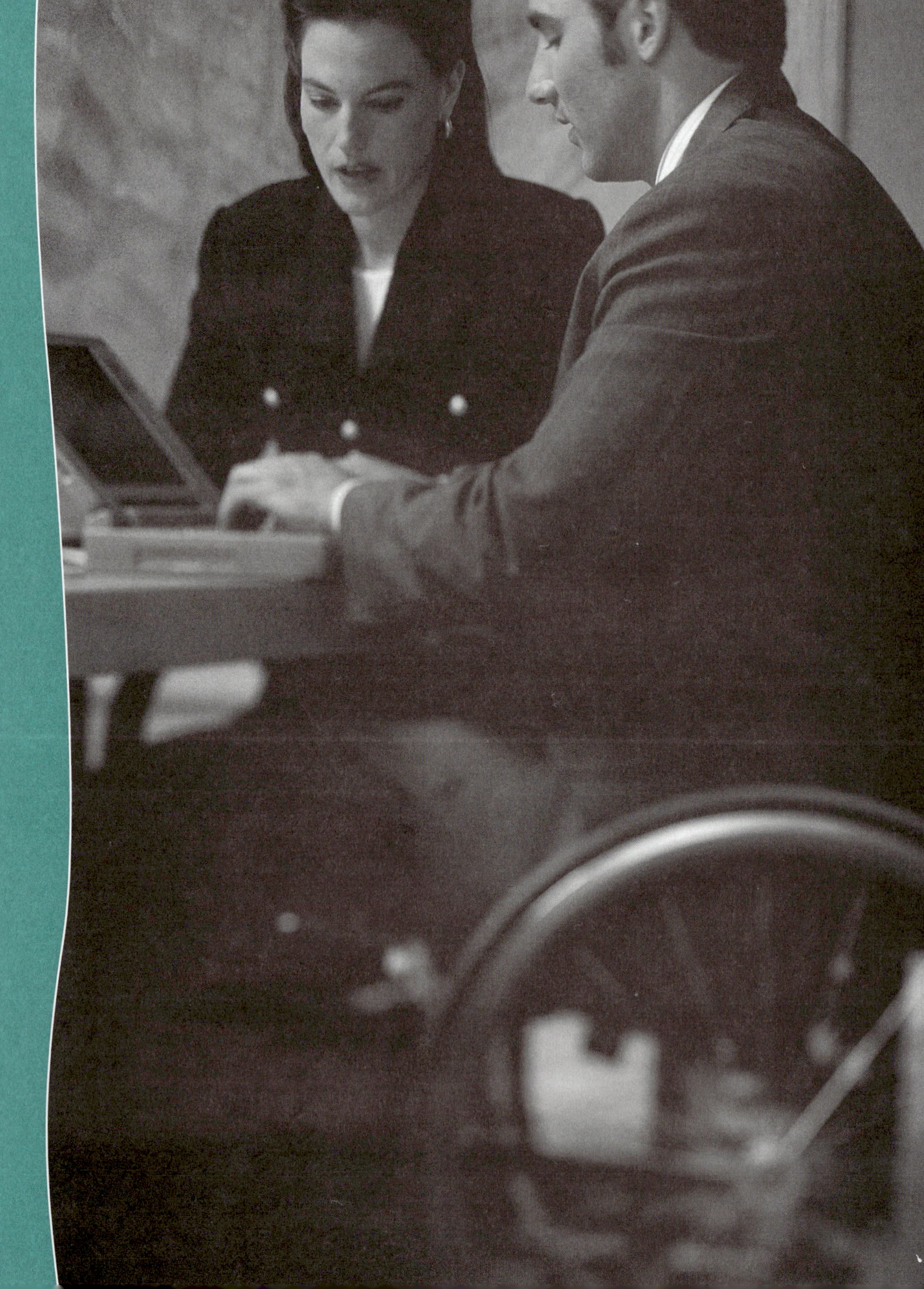

6 Getting the Process Started

Once you've selected the college of your choice, you will have to complete a series of steps before you can actually attend. This process may seem like a giant maze with one hurdle after another, but getting organized and understanding the steps will help you accomplish your goal.

In this chapter, you will explore answers to the following questions:

- What are the common admission requirements?
- How do I complete the admission process in an organized manner?
- What does early admission mean, and what are its advantages?
- What do I need to know about financial aid?
- How will I register for classes?

WHAT ARE THE COMMON ADMISSION REQUIREMENTS?

One of the first steps you need to take is to apply for admission. Most colleges require similar information before admitting you, but it is important to find out exactly what your college requires so that your admission process is smooth and expedient.

Admission Definitions

Open admission
Anyone with the appropriate credentials will automatically be accepted.

Competitive admission
Students must meet higher than average admission standards, and the number of openings may be limited.

Colleges offer one of two types of admission: open and competitive. **Open admission** means that the college will accept any incoming freshman who has earned a high school diploma or GED and who has placed within the required range of scores for tests such as the American College Testing Program (ACT) or the Scholastic Assessment Test (SAT). Students with low test scores or GPAs may be admitted on a provisional status until they successfully complete developmental courses that will increase their skill level, or they may be directed to attend a community college to take developmental courses there. **Competitive admission** means that the college demands specific requirements before admitting a student. Those requirements might mean a higher-than-average GPA, a high class ranking, or recommendations from professionals in the field.

Commonly, colleges have open admission but competitive admission within specific programs. For example, a college may have open admission for freshmen, but when a student completes the sophomore year, that student may have to apply to enter a particular program such as social work or education.

Minimum Required Information for Admission

Even colleges with open admission policies require a record of your past academic performance. You should begin a permanent file that contains the following documents:

- High school transcripts and documentation of grade point average
- College transcripts if you've taken courses while still in high school
- Documentation of class ranking (usually found on transcript)
- Documentation of ACT or SAT scores

Keep this file current and in a convenient place so that your documents are easily accessible if you want to apply to more than one college.

Transcripts

Transcripts are a permanent list of classes and the grades you've earned in those classes. High school transcripts may also contain information about overall grade point average, attendance, and class ranking. College transcripts will list all classes you enroll in and the grades you earn. It will list classes you withdraw from and audit, as well. Grade point averages, earned degrees, and graduation honors will also be listed on college transcripts.

High School Grade Point Averages

Even colleges with open admission policies demand that students have completed a precollege curriculum and have earned a GPA that meets their minimum standards. This baseline varies from college to college, so research your college's admission standards to see if you qualify.

A somewhat common GPA minimum standard is 2.5; however, if students don't have a 2.5 GPA or if they've earned a GED, a college may accept that if the student has earned a higher-than-minimum score on ACT or SAT composite scores.

Precollege Curriculum

Preparatory curriculum varies from state to state, but, in general, colleges with open admission policies insist that incoming college students have completed specific requirements in the core academic areas. Commonly, those requirements include completing four units of English, three units of math, three units of social science, three units of natural science, and two units of foreign language. If you are nearing graduation and haven't completed a precollege core of classes, you might want to consider summer school.

Tests

As part of their admission process, colleges generally require the scores of a standardized test. The two tests that are most common are the American College Testing Program and the Scholastic Assessment Test. The scores of these tests are used differently by colleges that have competitive admission than by colleges with open admission.

Colleges with competitive admission use these scores as one means of selecting students. Students with high ACT or SAT scores may be accepted to a number of colleges while students with average to low ACT or SAT scores may have difficulty getting accepted to schools with competitive admission.

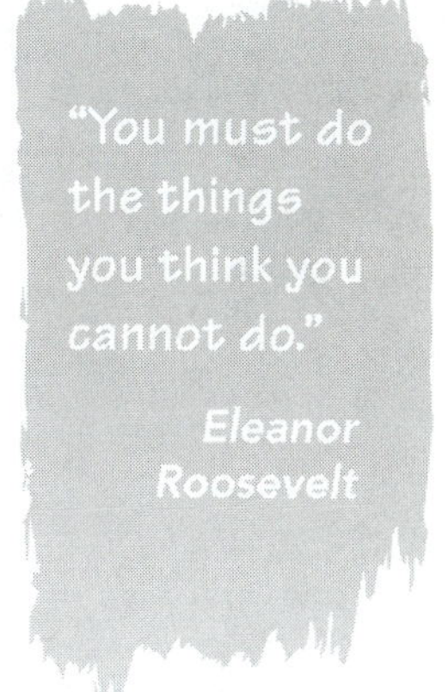

Colleges with open admission use ACT and SAT scores to determine if students meet basic academic competency. If a student scores low in specific areas, that student may be admitted on a provisional basis until the deficiency can be corrected by taking basic developmental courses.

Occasionally community colleges will not require that you submit ACT or SAT scores; however, these colleges will require that you take a placement test at the college. These scores are used to place students in courses that are best suited for their academic abilities. If the college you are considering requires that you complete a placement test, make sure you know when and where you take the test because these tests are required before you can enroll in classes.

Students who have completed precollege curriculum, earned high GPAs, and scored in the above-average to high range on the ACT or SAT may want to consider taking the College Level Examination Program (CLEP) test. CLEP tests will determine whether a student has college-level knowledge about a particular subject. When a student "CLEPs" out of a class, this means the student will get credit and will pay for the class but will not have to actually take the course.

Although tests are an important part of the admission process, admissions counselors understand that test scores are only one indicator of how well a student may do. If you have lower-than-expected test scores, you should emphasize other strengths you have as a student.

HOW DO I COMPLETE THE ADMISSION PROCESS IN AN ORGANIZED MANNER?

Even though colleges require the same general information for admission applications, there is a great deal of information to keep organized. Starting a filing system early will help you through the process. We suggest you keep separate files for copies of all the general information we've discussed.

In each file, keep three or four copies of each document, and label the original so you don't accidentally mail it. Most colleges only require photocopies of documents until admission has been approved. At that time, colleges can request that you send official transcripts. Sending copies will save you a great deal of money if you are applying to several colleges.

Also keep a file that contains a copy of your admission application for each college you apply to. Attach to each application a list of all documents you have submitted. When you have received notice of your admission status, place that notice in your file until you have made your decision.

When you apply for admission to a college, do so in an organized manner (see Figure 6.1) to make a good first impression:

- ✓ Write a cover letter that discusses required information for competitive admission colleges.
- ✓ Complete every question on the application.
- ✓ Attach all required documents in order.

Some colleges provide online admission applications on their Web pages. If you choose to apply electronically, don't forget to follow up with the appropriate documents.

FIGURE 6.1 Have I . . .

APPLIED FOR ADMISSION?

- Obtain admission application through your counselor, directly from the college, or from its online resource.
- Obtain copies of high school transcripts.
- Obtain copies of test scores, such as ACT or SAT.
- Complete and mail application and required documents.

WHAT DOES EARLY ADMISSION MEAN, AND WHAT ARE ITS ADVANTAGES?

Early admission has two different meanings. In some cases, high school students can apply for early admission to a college and take classes while still in high school. There are often specific requirements for this type of early admission that may include a specific grade point average, an interview process, and referrals from high school officials. Clearly this type of early admission is advantageous because it allows students to get a feel for college to see if they would like to attend there after high school. It is also a way to complete general education requirements and take time to explore personal interests.

The other definition of early admission is simply completing the admission process early in the year prior to attending. For example, students planning to attend college in the fall may complete an early admission process in the spring or early summer. This type of early admission also has its advantages. Besides having a larger selection of classes from which to choose, a student who applies early may also be able to take advantage of special orientations or introductory sessions. These orientations may give students one-to-one mentoring, a stay in campus housing, special advising sessions, and social time to meet other new students.

WHAT DO I NEED TO KNOW ABOUT FINANCIAL AID?

While you are in the process of applying for admission to the colleges that you are considering, you should also apply for financial aid. Seeking help from various sources of financial aid has become a way of life for much of the student population. Education is an important but often expensive investment. The cost for a year's full-time tuition only (not including room and board) in 1995–1996 ranged from $900 to $15,000, with the national average hovering around $2,100 for public institutions and over $11,000 for private ones.[1]

Not many people can pay for tuition in full without aid. In fact, almost half of students enrolled receive some kind of aid.[2]

Most sources of financial aid don't seek out recipients. Take the initiative to learn how you (or you and your parents, if they currently help to support you) can finance your education. Find the people on campus who can help you with your finances. Do some research to find out what's available, weigh the pros and cons of each option, and decide what would work best for you. Try to apply as early as you can. The types of financial aid available to you are loans, grants, and scholarships.

Loans

A loan is given to you by a person, bank, or other lending agency, usually to put toward a specific purchase. You, as the recipient of the loan, then must pay back the amount of the loan, plus interest, in regular payments that stretch over a particular period of time. Interest is the fee that you pay for the privilege of using money that belongs to someone else.

Loan Applications

What happens when you apply for a loan?

1. The loaning agency must approve you. You (and your parents) may be asked about what you (and any other family members) earn, how much

savings you have, your credit history, anything you own that is of substantial value (e.g., a car), and your history of payment on any previous loans.

2. An interest charge will be set. Interest can range from 5 percent to over 20 percent, depending on the loan and the economy. Variable-interest loans shift charges as the economy strengthens or weakens. Fixed-rate loans have one interest rate that remains constant.

3. The loaning agency will establish a payment plan. Most loan payments are made monthly or quarterly (four times per year). The payment amount depends on the total amount of the loan, how much you can comfortably pay per month, and the length of the repayment period.

Types of Student Loans

The federal government administers or oversees most student loans. To receive aid from any federal program, you must be a citizen or eligible noncitizen and be enrolled in a program of study that the government has determined is eligible. Individual states may differ in their aid programs. Check with the financial aid office of the colleges you apply to to find out details about your state and those colleges in particular.

Following are the main student loan programs to which you can apply if you are eligible. Amounts vary according to individual circumstances. Contact your school or federal student aid office for further information. In most cases, the amount is limited to the cost of your education minus any other financial aid you are receiving. All the following information on federal loans and grants comes from *The 2000–2001 Student Guide to Financial Aid,* published by the U.S. Department of Education[3]:

Perkins loans. Carrying a low, fixed rate of interest, these loans are available to those with exceptional financial need (need is determined by a government formula that indicates how large a contribution toward your education your family should be able to make). Schools issue these loans from their own allotment of federal education funds. After you graduate, you have a grace period (up to nine months, depending on whether you were a part-time or full-time student) before you have to begin repaying your loan in monthly installments.

Stafford loans. Students enrolled in school at least half-time may apply for a Stafford loan. Exceptional need is not required. However, students who can prove exceptional need may qualify for a subsidized Stafford loan, for which the government pays your interest until you begin repayment. There are two types of Stafford loans. A direct Stafford loan comes from government funds, and an FFEL (Federal Family Education Loan) Stafford loan comes from a bank or credit union participating in the FFEL program. The type available to you depends on your school's financial aid program. You begin to repay a Stafford loan six months after you graduate, leave school, or drop below half-time enrollment.

Plus loans. Your parents can apply for a Plus loan if they currently claim you as a dependent and if you are enrolled at least half-time. They must also undergo a credit check in order to be eligible, although the loans are not based on income. If they do not pass the credit check, they may be able to sponsor the loan through a relative or friend who does pass. Interest is variable; the loans are available from either the government or banks and credit unions. Your parents will have to begin repayment sixty days after they receive the last loan payment; there is no grace period.

For a few students, a loan from a relative is possible. If you have a close relationship with a relative who has some money put away, you might be able to talk to that person about helping you with your education. Discuss the terms of the loan as you would with any financial institution, detailing how and when you will receive the loan as well as how and when you will repay it. It may help to put the loan in writing. You may want to show your gratitude by offering to pay interest.

Grants and Scholarships

Both grants and scholarships require no repayment and therefore give your finances a terrific boost. Grants, funded by the government, are awarded to students who show financial need. Scholarships are awarded to students who show talent or ability in the area specified by the scholarship. They may be financed by government or private organizations, schools, or individuals.

Federal Grant Programs

Pell grants. These grants are need-based. The Department of Education uses a standard formula to evaluate the financial information you report on your application and determines your eligibility from that score (called an EFC, or expected family contribution, number). You must also be an undergraduate student who has earned no other degrees to be eligible. The Pell grant serves as a foundation of aid to which you may add other aid sources, and the amount of the grant varies according to the cost of your education and your EFC. Pell grants require no repayment.

Federal Supplemental Educational Opportunity Grants (FSEOG). Administered by the financial aid administrator at participating schools, SEOG eligibility depends on need. Whereas the government guarantees that every student eligible for a Pell grant will receive one, each school receives a limited amount of federal funds for SEOGs, and once it's gone, it's gone. Schools set their own application deadlines. Apply early. No repayment is required.

Work-study. Although you work in exchange for the aid, work-study is considered a grant because a limited number of positions are available. This program is need-based and encourages community service work or work related in some way to your course of study. You will earn at least the federal minimum wage and will be paid hourly. Jobs can be on campus (usually for your school) or off campus (often with a nonprofit organization or a local, state, or federal public agency). Find out who is in charge of the work-study program at the colleges where you apply.

There is much more to say about these financial aid opportunities than can be discussed here. Many other important details about federal grants and loans are available in *The 2000–2001 Student Guide to Financial Aid.* You might find this information at a college financial aid office, or you can request it by mail, phone, or online service:

Address: Federal Student Aid Information Center
P.O. Box 84
Washington, D.C. 20044

Phone: 1-800-4-FED-AID (1-800-433-3243)
TDD for the hearing-impaired: 1-800-730-8913

Web site: www.ed.gov/prog_info/SFA/StudentGuide

Scholarships

Scholarships are given for different kinds of abilities and talents: Some reward academic achievement, some reward exceptional abilities in sports or the arts, and some reward citizenship or leadership. Certain scholarships are sponsored by federal agencies. If you display exceptional ability and are disabled, are female, have an ethnic background classified as a minority (such as African American or Native American), or are a child of someone who draws benefits from a state agency (such as a POW or MIA), you might find scholarship opportunities geared toward you.

All kinds of organizations offer scholarships. You may receive scholarships from individual departments at your school or your school's independent scholarship funds, local organizations such as the Rotary Club, or privately operated aid foundations. Labor unions and companies may offer scholarship opportunities for children of their employees. Membership groups such as scouting organizations or the Y might offer scholarships, and religious organizations such as the Knights of Columbus or the Council of Jewish Federations might be another source.

Sources for Grants and Scholarships

It can take work to locate grants, scholarships, and work-study programs because many of them aren't widely advertised. Ask at your school's guidance office or a college's financial aid office. Visit your library or bookstore and look in the section on college or financial aid. Guides to funding sources, such as Richard Black's *The Complete Family Guide to College Financial Aid* and others, catalog thousands of organizations and help you find what fits you. Check out online scholarship search services. Use common sense and time management when applying for aid—fill out the application as neatly as possible, and send it in on time or even early. In addition, be wary of scholarship scam artists who ask you to pay a fee up front for them to find aid for you.

After you have completed the financial aid process (see Figure 6.2) and have decided which college to attend, you will register for classes.

HOW WILL I REGISTER FOR CLASSES?

After you have been accepted to the college you will attend, you will need to register for classes. Even though many colleges allow students to register for classes online, you should set up an appointment with an advisor the first

Have I . . .

FIGURE 6.2

MADE FINANCIAL ARRANGEMENTS?

- Obtain financial aid forms from your high school counselor or directly from the college you are applying to.
- Completely fill out forms, sign the forms, and mail to the appropriate address.
- Determine how to apply for scholarships, and follow through on the instructions.
- Apply for part-time on-campus jobs if necessary.
- Apply for bank loans if necessary.
- Apply for and put down appropriate deposit or down payment for residential halls or apartment.
- Check on fees for other expenses such as meal plans, parking, activities, and insurance.

time so that you have a clear understanding of the classes you will need to complete in order to earn a degree. In fact, the safest move is to meet with your advisor every semester so that your progress will be monitored.

When you meet with your advisor, you will receive a degree plan. This is a list of courses you will be required to successfully complete in order to graduate with a specific degree. Keep this list! A degree plan acts as a legal document between you and the college. Should the college decide to change the degree plan before you graduate, you probably will not be required to take additional classes if you are clearly progressing on an approved degree plan.

Once you have registered for classes (see Figure 6.3), you will be on your way! There may be other decisions you will need to make, including housing, meals, and transportation. These, too, are important decisions and will have an impact on your college experience, so work through these decisions carefully.

When you begin your college experience, continue to evaluate and refine your personal mission statement and your long- and short-term goals, as well as your personal skills and study skills. These are skills you can take with you on your journey of lifelong learning.

Have I . . .

FIGURE 6.3

REGISTERED FOR CLASSES?

- Meet with an advisor (this may be a faculty member or a staff member who works in student services) to determine which classes you will enroll in.
- Create a class schedule that will be based on your academic needs, as well as your personal needs. Take into consideration extracurricular activities or jobs you might be involved with.
- Take registration documents to the appropriate office.
- Pay registration fees or a down payment to hold your classes.

EXERCISES AND ACTIVITIES

6.1 Admission Requirements

Gather college catalogs from colleges you are considering. Research the admission requirements and assemble a packet of required materials.

6.2 Financial Aid Applications

Obtain current financial aid applications to fill out. With your parents, complete the forms and send them to the colleges you are considering.

6.3 Student Loan Calculations

Research how to take out a student loan. Calculate the cost of this loan, based on current interest rates. Determine if this is an option you want to consider. If not, use the problem-solving model found in Chapter 2 to determine how you will meet college expenses without this type of financial aid.

6

Endnotes

1 Thomas D. Snyder, Charlene M. Hoffman, and Claire M. Geddes, U.S. Department of Education, National Center for Education Statistics, *Digest of Education Statistics 1996*, NCES 96-133 (Washington, DC: U.S. Government Printing Office, 1996), 320–321.

2 U.S. Department of Education, *The 2000–2001 Student Guide to Financial Aid.*

3 Ibid.

Part III Keys to College Success

Study Skills to Master for Success

Chapter 7 Reading

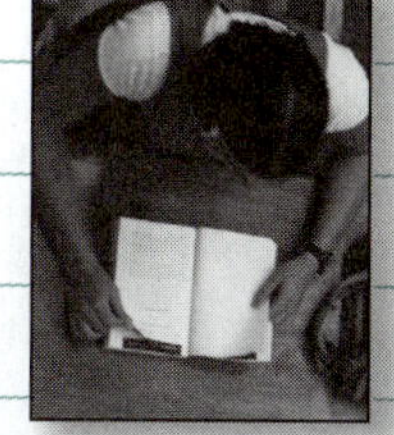

Chapter 8 Listening and Note Taking

Chapter 9 Test Taking

Personal Skills to Master for Success

Chapter 10 Managing Time

Chapter 11 Communicating

Chapter 12 Being a Leader

L E A R N

LEARN

7

Reading

The society you live in revolves around the written word. Although the growth of computer technology may seem to have made technical knowledge more important than reading, the focus on word processing and computer handling of documents has actually increased the need for employees who function at a high level of literacy. As *The Condition of Education 1996* report states, "In recent years, literacy has been viewed as one of the fundamental tools necessary for successful economic performance in industrialized societies. Literacy is no longer defined merely as a basic threshold of reading ability, but rather as the ability to understand and use printed information in daily activities, at home, at work, and in the community."[1]

Two crucial keys to your college success are reading and studying. If you read thoroughly and understand what you read, and if you achieve your study goals, you can improve your capacity to learn and understand. In this chapter, you will learn how you can overcome barriers to successful reading and benefit from defining a purpose each time you read. You will explore the PQ3R study technique and see how critical reading can help you maximize your understanding of any text.

In this chapter, you will explore answers to the following questions:

- What are some challenges of reading?
- Why define my purpose for reading?
- How can PQ3R help me study?
- How can I read critically?

WHAT ARE SOME CHALLENGES OF READING?

Whatever your skill level, you will encounter challenges that make reading more difficult, such as an excess of reading assignments, difficult texts, distractions, a lack of speed and comprehension, and insufficient vocabulary. Following are some ideas about how to meet these challenges. Note that if you have a reading disability, if English is not your primary language, or if you have limited reading skills, you may need additional support and guidance. Most colleges provide services for students through a reading center or tutoring program. Take the initiative to seek help if you need it. Many accomplished learners have benefited from help in specific areas.

Dealing with Reading Overload

Reading overload is part of almost every college experience. On a typical day, you may be faced with reading assignments that look like this:

- An entire textbook chapter on the causes of the Civil War (American history)
- An original research study on the stages of sleep (psychology)
- Pages 1–50 in Arthur Miller's play *Death of a Salesman* (American literature)

Reading all this and more leaves little time for anything else unless you read selectively and skillfully. You can't control your reading load. You can, however, improve your reading skills. The material in this chapter will present techniques that can help you read and study as efficiently as you possibly can—and still have time left over for other things.

Working Through Difficult Texts

Although many textbooks are useful teaching tools, some can be poorly written and organized. Students using texts that aren't well written may blame themselves for the difficulty they're experiencing. Because texts are often written with the purpose of challenging the intellect, even well-written and organized texts may be difficult and dense to read. Generally, the further you advance in your education, the more complex your required reading is likely to be. For example, your sociology professor may assign a chapter on the dynamics of social groups, including those of dyads and triads. When is the last time you heard the terms dyads and triads in normal conversation? You may feel at times as though you are reading a foreign language as you encounter new concepts, words, and terms.

Making your way through poorly written or difficult reading material is hard work that can be accomplished through focus, motivation, commitment, and skill. The following strategies may help:

- Approach your reading assignments head-on. Be careful not to prejudge them as impossible or boring before you even start to read.
- Accept the fact that some texts may require some extra work and concentration. Set a goal to make your way through the material and learn, whatever it takes.
- When a primary source discusses difficult concepts that it does not explain, put in some extra work to define such concepts on your own. Ask your instructor or other students for help. Consult reference materials in that particular subject area, other class materials, dictionaries, and encyclopedias. You may want to make this process more convenient by creating your own mini-library at home. Collect reference materials that you use often, such as a dictionary, a thesaurus, a writer's style handbook, and maybe an atlas or computer manual. You may also benefit from owning reference materials in your particular areas of study. "If you find yourself going to the library to look up the same reference again and again, consider purchasing that book for your personal or office library," advises library expert Sherwood Harris.[2]
- Look for order and meaning in seemingly chaotic reading materials. The information you will find in this chapter on the PQ3R reading technique and on critical reading will help you discover patterns and achieve a greater depth of understanding. Finding order within chaos is an important skill, not just in the mastery of reading but also in life. This skill can give you power by helping you "read" (think through) work dilemmas, personal problems, and educational situations.

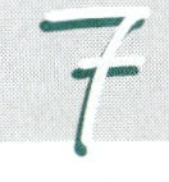

Managing Distractions

With so much happening around you, it's often hard to keep your mind on what you are reading. Distractions take many forms. Some are external: the sound of a telephone, a friend who sits next to you at lunch and wants to talk, a sibling who asks for help with homework. Other distractions come from within. As you try to study, you may be thinking about the game on Friday night, an argument you had with a friend, a paper due in art history, or a site on the Internet you want to visit.

Identify the distraction and choose a suitable action. Pinpoint what's distracting you before you decide what kind of action to take. If the distraction is external and out of your control, such as construction outside your building or a noisy group in the library, try to move away from it. If the distraction is external but within your control, such as the television, telephone, or siblings, take action. For example, if the TV or phone is a problem, turn off the TV or unplug the phone for an hour.

If the distraction is internal, there are a few strategies to try that may help you clear your mind. You may want to take a break from your studying and tend to one of the issues that you are worrying about. Physical exercise may relax you and bring back your ability to focus. For some people, studying while listening to music helps to quiet a busy mind. For others, silence may do the trick. If you need silence to read or study and cannot find a truly quiet environment, consider purchasing sound-muffling headphones or even earplugs.

Find the best place and time to read. Any reader needs focus and discipline in order to concentrate on the material. Finding a place and time that minimize outside distractions will help you achieve that focus. Here are some suggestions. Read alone, unless you are working with other readers. Family members, friends, or others who are not in the study mode may interrupt your concentration. If you prefer to read alone, establish a relatively interruption-proof place and time, such as an out-of-the-way spot at the library or an after-class time in an empty classroom. If you study at home, you may want to place a "Quiet" sign on the door. Some students benefit from reading with one or more other students. If this helps you, plan to schedule a group reading meeting where you read sections of the assigned material and then break to discuss them.

Find a comfortable location. Many students study in the library on a hard-backed chair. Others prefer a library easy chair, a chair in their room, or even the floor. The spot you choose should be comfortable enough for hours of reading, but not so comfortable that you fall asleep. Also, make sure that you have adequate lighting and aren't too hot or too cold.

Choose a regular reading place and time. Choose a spot or two you like and return to them often. Also, choose a time when your mind is alert and focused. Some students prefer to read just before or after the class for which the reading is assigned. Eventually, you will associate preferred places and times with focused reading.

If it helps you concentrate, listen to soothing background music. The right music can drown out background noises and relax you. However, the wrong music can make it impossible to concentrate; for some people, silence is better. Experiment to learn what you prefer; if music helps, stick with the type that works best. A personal headset makes listening possible no matter where you are.

Note: Turn off the television. For most people, reading and TV don't mix.

Building Comprehension and Speed

Most students lead busy lives, carrying heavy academic loads while perhaps working a part-time job and participating in extracurricular activities. It's difficult to make time to study at all, let alone handle the enormous reading assignments for your different classes. Increasing your reading comprehension and speed will save you valuable time and effort.

Rapid reading won't do you any good if you can't remember the material or answer questions about it. However, reading too slowly can be equally inefficient because it often eats up valuable study time and gives your mind space to wander. Your goal is to read for maximum speed and comprehension. Focus on comprehension first because greater comprehension is the primary goal and also promotes greater speed.

Following are some specific strategies for increasing your understanding of what you read:

METHODS FOR INCREASING READING COMPREHENSION

- Continually build your knowledge through reading and studying. More than any other factor, what you already know before you read a passage will determine your ability to understand and remember important ideas. Previous knowledge, including vocabulary, facts, and ideas, gives you a **context** for what you read.
- Establish your purpose for reading. When you establish what you want to get out of your reading, you will be able to determine what level of understanding you need to reach and, therefore, on what you need to focus.
- Remove the barriers of negative self-talk. Instead of telling yourself that you cannot understand, think positively. Tell yourself, I can learn this material. I am a good reader.
- Think critically. Ask yourself questions: Do I understand the sentence, paragraph, or chapter I just read? Are ideas and supporting examples clear to me? Could I clearly explain to someone else what I just read?

Context
Written or spoken knowledge that can help to illuminate the meaning of a word or passage.

The following suggestions will help increase your reading speed:

METHODS FOR INCREASING READING SPEED

- Try to read groups of words rather than single words.
- Avoid pointing your finger to guide your reading because this will slow your pace.
- Try swinging your eyes from side to side as you read a passage, instead of stopping at various points to read individual words.
- When reading narrow columns, focus your eyes in the middle of the column and read down the page. With practice, you'll be able to read the entire column width.
- Avoid **vocalization** when reading.
- Avoid thinking each word to yourself as you read it, a practice known as subvocalization. Subvocalization is one of the primary causes of slow reading speed.

Vocalization
The practice of speaking the words and/or moving your lips while reading.

Facing the challenges of reading is only the first step. The next important step is to examine why you are reading any given piece of material.

WHY DEFINE MY PURPOSE FOR READING?

As with all other aspects of your education, asking important questions will enable you to make the most of your efforts. When you define your purpose, you ask yourself why you are reading a particular piece of material. One way to do this is by completing this sentence: "In reading this material, I intend to define/learn/answer/achieve . . . " With a clear purpose in mind, you can decide how much time and what kind of effort to expend on various reading assignments. Nearly 375 years ago, Francis Bacon, the English philosopher, stated:

> Some books are to be tasted, others to be swallowed, and some few to be chewed and digested; that is, some books are to be read only in parts, others to be read but not curiously; and some few to be read wholly, and with diligence and attention.

"In books, I could travel anywhere, be anybody, understand worlds long past and imaginary colonies in the future."
Rita Dove

Achieving your reading purpose requires adapting to different types of reading materials. Being a flexible reader—adjusting your reading strategies and pace—will help you to adapt successfully.

Purpose Determines Reading Strategy

With purpose comes direction; with direction comes a strategy for reading. Following are four reading purposes, examined briefly. You may have one or more for each piece of reading material you approach:

Purpose 1: Read to evaluate critically. Critical evaluation involves approaching the material with an open mind, examining causes and effects, evaluating ideas, and asking questions that test the strength of the writer's argument and that try to identify assumptions. Critical reading is essential for you to demonstrate an understanding of material that goes beyond basic recall of information. You will read more about critical reading later in the chapter.

Purpose 2: Read for comprehension. Much of the studying you do involves reading for the purpose of comprehending the material. The two main components of comprehension are general ideas and specific facts/examples. These components depend on one another. Facts and examples help to explain or support ideas, and ideas provide a framework that helps the reader to remember facts and examples:

- *General ideas.* General-idea reading is rapid reading that seeks an overview of the material. You may skip entire sections as you focus on headings, subheadings, and summary statements in search of general ideas.
- *Specific facts/examples.* At times, readers may focus on locating specific pieces of information—for example, the stages of intellectual development in young children. Often, a reader may search for examples that support or explain more general ideas—for example, the causes of economic recession. Because you know exactly what you are looking for, you can skim the material at a rapid rate. Reading your texts for specific information may help before taking a test.

Purpose 3: Read for practical application. A third purpose for reading is to gather usable information that you can apply toward a specific goal. When you read a computer software manual, an instruction sheet for assembling a gas grill, or a cookbook recipe, your goal is to learn how to do something. Reading and action usually go hand in hand.

Purpose 4: Read for pleasure. Some materials you read for entertainment, such as a *Sports Illustrated* magazine or the latest John Grisham courtroom thriller. Entertaining reading may also go beyond materials that seem obviously designed to entertain. Whereas some people may read a Jane Austen novel for comprehension, as in a class assignment, others may read Austen books for pleasure.

Use PQ3R to become an active reader.

ACTIVE READERS TEND TO . . .	
Divide material into manageable sections	Answer end-of-chapter questions and applications
Write questions	Create chapter outlines
Answer questions through focused note taking	Create think links that map concepts in a logical way
Recite, verbally and in writing, the answers to questions	Make flash cards and study them
Highlight key concepts	Recite what they learned into a tape recorder and play the tape back
Focus on main ideas found in paragraphs, sections, and chapters	Rewrite and summarize notes and highlighted materials from memory
Recognize summary and support devices	Explain what they read to a family member or friend
Analyze tables, figures, and photos	Form a study group

Purpose Determines Pace

George M. Usova, senior education specialist and graduate professor at Johns Hopkins University, explains: "Good readers are flexible readers. They read at a variety of rates and adapt them to the reading purpose at hand, the difficulty of the material, and their familiarity with the subject area."[3] As Table 7.1 shows, good readers link the pace of reading to their reading purpose.

7

HOW CAN PQ3R HELP ME STUDY READING MATERIAL?

Reading also gives you mastery over concepts. For example, a biology student learns what happens during photosynthesis, and a business student learns about marketing research.

This section will focus on a technique that will help you learn and study more effectively as you read your college textbooks.

Preview-Question-Read-Recite-Review (PQ3R)

PQ3R is a technique that will help you grasp ideas quickly, remember more, and review effectively and efficiently for tests. The symbols P-Q-3-R stand for preview, question, read, recite, and review—all steps in the studying process. Developed more than 55 years ago by Francis Robinson,[4] the technique is still being used today because it works. It is particularly helpful for studying texts. When reading literature, read the work once from beginning to end to appreciate the story and language. Then, reread it using PQ3R to master the material.

Skimming
Rapid, superficial reading of material that involves glancing through to determine central ideas and main elements.

Scanning
Reading material in an investigative way, searching for specific information.

Moving through the stages of PQ3R requires that you know how to skim and scan. **Skimming** involves rapid reading of various chapter elements, including introductions, conclusions, and summaries; the first and last lines of paragraphs; boldface or italicized terms; pictures, charts, and diagrams. In contrast, **scanning** involves the careful search for specific facts and examples. You will probably use scanning during the review phase of PQ3R when you need to locate and remind yourself of particular information. In a chemistry text, for example, you may scan for examples of how to apply a particular formula.

Preview

The best way to ruin a whodunit novel is to flip through the pages to find out how everything turned out. However, when reading textbooks, previewing can help you learn and is encouraged. Previewing refers to the process of surveying, or prereading, a book before you actually study it. Most textbooks include devices that give students an overview of the text as a whole as well as the contents of individual chapters. As you look at Figure 7.1, think about how many of these devices you already use.

Question

Your next step is to examine the chapter headings and, on your own paper, write questions linked to those headings. These questions will focus your attention and increase your interest, helping you relate new ideas to what you already know and building your comprehension. You can take questions from the textbook or from your lecture notes or come up with them on your own when you preview, based on what ideas you think are most important.

FIGURE 7.1 Text and chapter previewing devices.

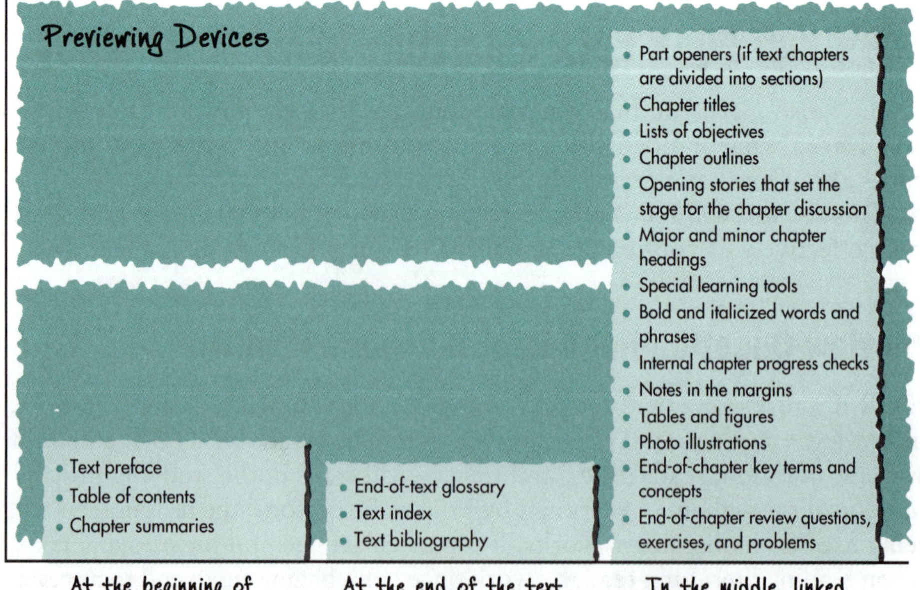

Formulating questions.

I. THE CONSUMER BUYING PROCESS	I. WHAT IS THE CONSUMER BUYING PROCESS?
A. Problem/Need Recognition	A. Why must consumers first recognize a problem or need before they buy a product?
B. Information Seeking	B. What is information seeking, and who answers consumers' questions?
C. Evaluation of Alternatives	C. How do consumers evaluate different products to narrow their choices?
D. Purchase Decision	D. Are purchasing decisions simple or complex?
E. Postpurchase Evaluations	E. What happens after the sale?

Here is how this technique works. Referring to Figure 7.2, the left column contains primary- and secondary-level headings from a section of *Business,* an introductory text by Ricky W. Griffin and Ronald J. Ebert; the column on the right rephrases these headings in question form.

There is no correct set of questions. Given the same headings, you would create your own particular set of questions. The more useful kinds of questions are ones that engage the critical-thinking mind actions and processes found in Chapter 2.

Read

Your questions give you a starting point for reading, the first R in PQ3R. Read the material with the purpose of answering each question you raised. Pay special attention to the first and last lines of every paragraph, which should tell you what the paragraph is about. As you read, record key words, phrases, and concepts in your notebook. Some students divide the notebook into two columns, writing questions on the left and answers on the right. This method, known as the Cornell note-taking system, is described in more detail in Chapter 8.

If you own the textbook, marking it up—in whatever ways you prefer—is a must. The notations that you make will help you to interact with the material and make sense of it. You may want to write notes in the margins, circle key ideas, or highlight key sections. Some people prefer to underline, although underlining adds more ink to the lines of text and may overwhelm your eye. Although writing in a textbook makes it difficult to sell it back to the school, the increased depth of understanding you can gain is worth the investment.

Highlighting may help you pinpoint material to review before an exam. Here are some additional tips on highlighting:

- Get in the habit of marking the text after you read the material. If you do it while you are reading, you may wind up marking less important passages.

- Highlight key terms and concepts. Mark the examples that explain and support important ideas. You might try highlighting ideas in one color and examples in another.
- Highlight figures and tables. They are especially important if they summarize text concepts.
- Avoid overmarking. A phrase or two is enough in most paragraphs. Set off long passages with brackets rather than marking every line.
- Write notes in the margins with a pen or pencil. Comments like "main point" and "important definition" will help you find key sections later on.
- Be careful not to mistake highlighting for learning. You will not necessarily learn what you highlight unless you review it carefully. You may benefit from writing the important information you have highlighted into your lecture notes.

One final step in the reading phase is to divide your reading into digestible segments. Many students read from one topic heading to the next, then stop. Pace your reading so that you understand as you go. If you find you are losing the thread of the ideas you are reading, you may want to try smaller segments, or you may need to take a break and come back to it later.

Recite

Once you finish reading a topic, stop and answer the questions you raised about it in the Q stage of PQ3R. You may decide to recite each answer aloud, silently speak the answers to yourself, tell the answers to another person as though you were teaching him or her, or write your ideas and answers in brief notes. Writing is often the most effective way to solidify what you have read.

After you finish one section, move on to the next. Then repeat the question-read-recite cycle until you complete the entire chapter. If you find yourself fumbling for thoughts during this process, it means that you do not yet "own" the ideas. Reread the section that's giving you trouble until you master its contents. Understanding each section as you go is crucial because the material in one section often forms a foundation for the next.

Review

Review soon after you finish a chapter. Here are some techniques for reviewing:

- Skim and reread your notes. Then try summarizing them from memory.
- Answer the text's end-of-chapter review, discussion, and application questions.
- Quiz yourself, using the questions you raised in the Q stage. If you can't answer one of your own or one of the text's questions, go back and scan the material for answers.
- Review and summarize in writing the sections and phrases you have highlighted.
- Create a chapter outline in standard outline form or think-link form.
- Reread the preface, headings, tables, and summary.

- Recite important concepts to yourself, or record important information on a cassette tape and play it on your car's tape deck or your Walkman.
- Make flash cards that have an idea or word on one side and examples, a definition, or other related information on the other. Test yourself.
- Think critically. Break ideas down into examples, consider similar or different concepts, recall important terms, evaluate ideas, and explore causes and effects.
- Make think links that show how important concepts relate to one another.

Remember that you can ask your instructor if you need help clarifying your reading material. Your instructors are important resources. Pinpoint the material you want to discuss, schedule a meeting with him or her during or after school hours, and come prepared with a list of questions. You may also want to ask what materials to focus on when you study for tests.

If possible, you should review both alone and with study groups. Reviewing in as many different ways as possible increases the likelihood of retention. Figure 7.3 shows some techniques that will help a study group maximize its time and efforts.

Repeating the review process renews and solidifies your knowledge. That is why it is important to set up regular review sessions—for example, once a week. As you review, remember that refreshing your knowledge is easier and faster than learning it the first time.

HOW CAN I READ CRITICALLY?

Your textbooks will often contain features that highlight important ideas and help you determine questions to ask while reading. As you advance in your education, however, many reading assignments will not be so clearly marked, especially if they are primary sources. You will need critical-reading skills in order to select the important ideas, identify examples that support them, and ask questions about the text without the aid of any special features or tools.

Critical reading enables you to consider reading material carefully, developing a thorough understanding of it through evaluation and analysis. A critical reader is able to discern what in a piece of reading material is true or useful, such as when using material as a source for an essay. A critical reader can also compare one piece of material to another and evaluate which makes more sense, which proves its thesis more successfully, or which is more useful for the reader's purpose.

Critical reading is reading that transcends taking in and regurgitating material. You can read critically by using PQ3R to get a basic idea of the material, asking questions based on the critical-thinking mind actions, and engaging your critical-thinking processes.

Use PQ3R to "Taste" Reading Material

Sylvan Barnet and Hugo Bedau, authors of *Critical Thinking, Reading, and Writing: A Brief Guide to Argument,* suggest that the active reading of PQ3R will help you form an initial idea of what a piece of reading material is all

FIGURE 7.3 Study group techniques.

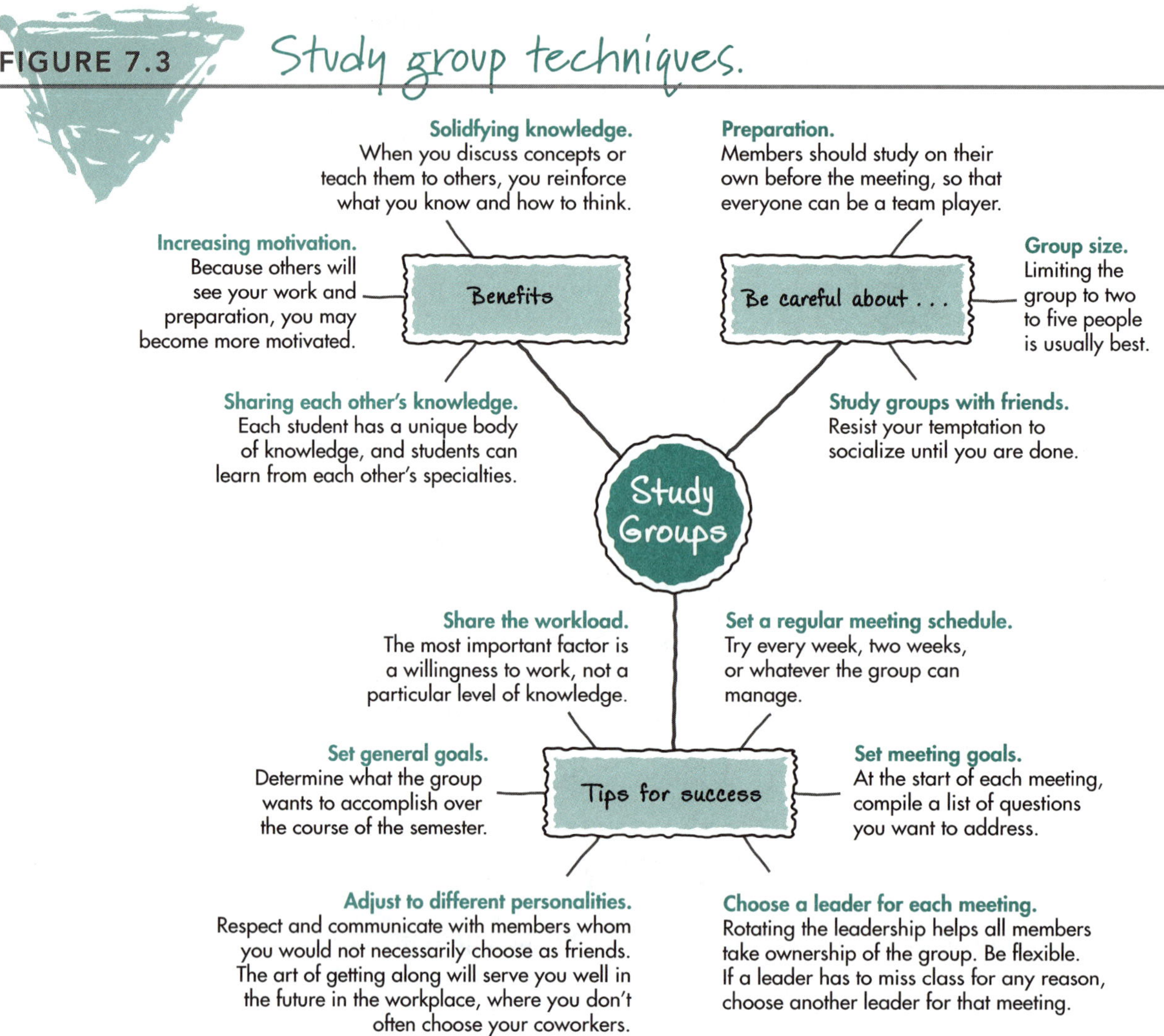

about. Through previewing, skimming for ideas and examples, highlighting and writing comments and questions in the margins, and reviewing, you can develop a basic understanding of its central ideas and contents.[5]

Summarizing, part of the review process in PQ3R, is one of the best ways to develop an understanding of a piece of reading material. To construct a **summary**, focus on the central ideas of the piece and the main examples that support those ideas. A summary does not contain any of your own ideas or your evaluation of the material. It simply condenses the material, making it easier for you to focus on the structure of the piece and its central ideas when you go back to read more critically. At that point, you can begin to evaluate the piece and introduce your own ideas. Using the mind actions of the Thinktrix (found in Chapter 2) will help you.

Summary
A concise restatement of the material, in your own words, that covers the main points.

Ask Questions Based on the Mind Actions

The essence of critical reading, as with critical thinking, is asking questions. Instead of simply accepting what you read, seek a more thorough understanding by questioning the material as you go along. Using the mind actions of the Thinktrix (found in Chapter 2) to formulate your questions will help you understand the material.

What parts of the material you focus on will depend on your purpose for reading. For example, if you are writing a paper on the causes of World War II, you might spend your time focusing on how certain causes fit your thesis. If you are comparing two pieces of writing that contain opposing arguments, you may focus on picking out their central ideas and evaluating how well the writers use examples to support these ideas.

You can question any of the following components of reading material:

- The central idea of the entire piece
- A particular idea or statement
- The examples that support an idea or statement
- The proof of a fact
- The definition of a concept

Following are some ways to critically question your reading material, based on the mind actions. Apply them to any component you want to question by substituting the component for the words "it" and "this."

Similarity:	What does this remind me of, or how is it similar to something else I know?
Difference:	What different conclusions are possible? How is this different from my experience?
Cause and effect:	Why did this happen, or what caused this? What are the effects or consequences of this? What effect does the author want to have, or what is the purpose of this material? What effects support a stated cause?
Example to idea:	How would I classify this, or what is the best idea to fit this example? How would I summarize this, or what are the key ideas? What is the thesis or central idea?
Idea to example:	What evidence supports this, or what examples fit this idea?
Evaluation:	How would I evaluate this? Is it valid or pertinent? Does this example support my thesis or central idea?

Analyze Perspective

Your understanding of perspective will help you understand that many reading materials are written from a particular perspective. Perspective often has a strong effect on how the material is presented. For example, if a recording artist and a music censorship advocate were to each write a piece about a controversial song created by that artist, their different perspectives would result in two very different pieces of writing.

To analyze perspective, ask questions like the following:

- What perspective is guiding this? What are the underlying ideas that influence this material?

- Who wrote this, and what may be the author's perspective? For example, a piece on a new drug written by an employee of the drug manufacturer may differ from a doctor's evaluation of the drug.
- What does the title of the material tell me about its perspective? For example, a piece titled "New Therapies for Diabetes" may be more informational, and "What's Wrong with Insulin Injections" may intend to be persuasive.
- How does the material's source affect its perspective? For example, an article on health management organizations (HMOs) published in an HMO newsletter may be more favorable and one-sided than one published in *The New York Times*.

Seek Understanding

Reading critically allows you to investigate what you read so that you can reach the highest possible level of understanding. Think of your reading process as an archaeological dig. The first step is to excavate a site and uncover the artifacts. In reading, that corresponds to your initial preview and reading of the material. As important as the excavation is, the process would be incomplete if you stopped there and just took home a bunch of items covered in dirt. The second half of the process is to investigate each item, evaluate what all of those items mean, and derive new knowledge and ideas from what you discover. Critical reading allows you to complete that crucial second half of the process.

As you work through all of the different requirements of critical reading, remember that critical reading takes time and focus. Finding a time, place, and purpose for reading, covered earlier in the chapter, is crucial to successful critical reading. Give yourself a chance to gain as much as possible from what you read.

EXERCISES AND ACTIVITIES

Reading Technique Application 7.1

The following is from the "Groups and Organizations" chapter in the sixth edition of John J. Macionis's *Sociology,* a Prentice Hall text.[6] Using what you learned in this chapter about study techniques, complete the questions following the selection.

SOCIAL GROUPS

Virtually everyone moves through life with a sense of belonging; this is the experience of group life. A **social group** refers to *two or more people who identify and interact with one another.* Human beings continually come together to form couples, families, circles of friends, neighborhoods, churches, businesses, clubs, and numerous large organizations. Whatever the form, groups encompass people with shared experiences, loyalties, and interests. In short, while maintaining their individuality, the members of social groups also think of themselves as a special "we."

Groups, Categories, and Crowds

People often use the term "group" imprecisely. We now distinguish the group from the similar concepts of category and crowd.

Category

A *category* refers to people who have some status in common. Women, single fathers, military recruits, homeowners, and Roman Catholics are all examples of categories.

Why are categories not considered groups? Simply because, while the individuals involved are aware that they are not the only ones to hold that particular status, the vast majority are strangers to one another.

Crowd

A *crowd* refers to a temporary cluster of individuals who may or may not interact at all. Students sitting together in a lecture hall do engage one another and share some common identity as college classmates; thus, such a crowd might be called a loosely formed group. By contrast, riders hurtling along on a subway train or bathers enjoying a summer day at the beach pay little attention to one another and amount to an anonymous aggregate of people. In general, then, crowds are too transitory and too impersonal to qualify as social groups.

The right circumstances, however, could turn a crowd into a group. People riding in a subway train that crashes under the city streets generally become keenly aware of their common plight and begin to help each other. Sometimes such extraordinary experiences become the basis for lasting relationships.

Primary and Secondary Groups

Acquaintances commonly greet one another with a smile and the simple phrase "Hi! How are you?" The response is usually a well-scripted "Just fine, thanks. How about you?" This answer, of course, is often more formal than truthful. In most cases, providing a detailed account of how you are *really* doing would prompt the other person to beat a hasty and awkward exit.

Sociologists classify social groups by measuring them against two ideal types based on members' level of genuine personal concern. This variation is the key to distinguish *primary* from *secondary* groups.

According to Charles Horton Cooley (1864–1929), who is introduced in the box, a **primary group** is *a small social group whose members share personal and enduring relationships.* Bound together by *primary relationships,* individuals in primary groups typically spend a great deal of time together, engage in a wide range of common activities, and feel that they know one another well. Although not without periodic conflict, members of primary groups display sincere concern for each other's welfare. The family is every society's most important primary group.

Cooley characterized these personal and tightly integrated groups as *primary* because they are among the first groups we experience in life. In addition, the family and early play groups also hold primary importance in the socialization process, shaping attitudes, behavior, and social identity.

1. Identify the headings on the page and the relationship among them. Which headings are primary-level headings, which are secondary, and which are tertiary (third-level heads)? Which heading serves as an umbrella for the rest?
2. What do the headings tell you about the content of the page?
3. Identify the boldface and italicized terms. How do boldface terms differ from italicized terms?
4. After reading the chapter headings, write three study questions. List the questions.
5. Note which key phrases and sentences you would highlight. Write short marginal notes to help you review the material at a later point.
6. After reading this page, list four key concepts that you will need to study.

7.2 Reading Purpose Application

Read the paragraphs on the following page on kinetic and potential energy and the first law of thermodynamics taken from *Life on Earth* by Teresa Audesirk and Gerald Audesirk.[7] When you have finished, answer the following questions.

Reading for critical evaluation. Evaluate the material by answering these questions:

1. Were the ideas clearly supported by examples? If you feel one or more were not supported, give an example.
2. Did the author make any assumptions that weren't examined? If so, name one or more.
3. Do you disagree with any part of the material? If so, which part, and why?
4. Do you have any suggestions for how the material could have been presented more effectively?

Reading for practical application. Imagine you have to give a presentation on this material the next time the class meets. On a separate sheet of paper, create an outline or think link that maps out the key elements you would discuss.

Reading for comprehension. Answer the following questions to determine the level of your comprehension:

1. Name the two types of energy.
2. Which one "stores" energy?
3. Can kinetic energy be turned into potential energy?
4. What is the term that describes the basic properties and behaviors of energy?

Among the fundamental characteristics of all living organisms is the ability to guide chemical reactions within their bodies along certain pathways. The chemical reactions serve many functions, depending on the nature of the organism: to synthesize the molecules that make up the organism's body, to reproduce, to move, even to think. Chemical reactions either require or release **energy**, which can be defined simply as *the capacity to do work,* including synthesizing molecules, moving things around, and generating heat and light. In this chapter we discuss the physical laws that govern energy flow in the universe, how energy flow in turn governs chemical reactions, and how the chemical reactions within living cells are controlled by the molecules of the cell itself. Chapters 7 and 8 focus on photosynthesis, the chief "port of entry" for energy into the biosphere, and glycolysis and cellular respiration, the most important sequences of chemical reactions that release energy.

ENERGY AND THE ABILITY TO DO WORK

As you learned in Chapter 2, there are two types of energy: **kinetic energy** and **potential energy.** Both types of energy may exist in many different forms. Kinetic energy, or *energy of movement,* includes light (movement of photons), heat (movement of molecules), electricity (movement of electrically charged particles), and movement of large objects. Potential energy, or *stored energy,* includes chemical energy stored in the bonds that hold atoms together in molecules, electrical energy stored in a battery, and positional energy stored in a diver poised to spring (Fig. 4-1). Under the right conditions, kinetic energy can be transformed into potential energy, and vice versa. For example, the diver converted kinetic energy of movement into potential energy of position when she climbed the ladder up to the platform; when she jumps off, the potential energy will be converted back into kinetic energy.

To understand how energy flow governs interactions among pieces of matter, we need to know two things: (1) the quantity of available energy and (2) the usefulness of the energy. These are the subjects of the laws of thermodynamics, which we will now examine.

The Laws of Thermodynamics Describe the Basic Properties of Energy

All interactions among pieces of matter are governed by the two **laws of thermodynamics,** physical principles that define the basic properties and behavior of energy. The laws of thermodynamics deal with "isolated systems," which are any parts of the universe that cannot exchange either matter or energy with any other parts. Probably no part of the universe is completely isolated from all possible exchange with every other part, but the concept of an isolated system is useful in thinking about energy flow.

The First Law of Thermodynamics States that Energy Can Be Neither Created nor Destroyed

The **first law of thermodynamics** states that within any isolated system, energy can be neither created nor destroyed, although it can be changed in form (for example, from chemical energy to heat energy). In other words, within an isolated system *the total quantity of energy remains constant.* The first law is therefore often called the law of conservation of energy. To use a familiar example, let's see how the first law applies to driving your car (Fig. 4-2). We can consider that your car (with a full tank of gas), the road, and the surrounding air roughly constitute an isolated system. When you drive your car, you convert the potential chemical energy of gasoline into kinetic energy of movement and heat energy. The total amount of energy that was in the gasoline before it was burned is the same as the total amount of this kinetic energy and heat.

An important rule of energy conversions is this: Energy always flows "downhill," from places with a high concentration of energy to places with a low concentration of energy. This is the principle behind engines. As we described in Chapter 2, temperature is a measure of how fast molecules move. The burning gasoline in your car's engine consists of molecules moving at extremely high speeds: a high concentration of energy. The cooler air outside the engine consists of molecules moving at much lower speeds: a low concentration of energy. The molecules in the engine hit the piston harder than the air molecules outside the engine do, so the piston moves upward, driving the gears that move the car. Work is done. When the engine is turned off, it cools down as heat is transferred from the warm engine to its cooler surroundings. The molecules on both sides of the piston move at the same speed, so the piston stays still. No work is done.

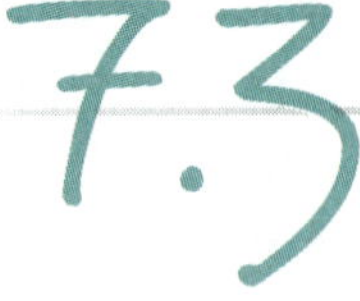

Research and Reading Technique Application

Find an article about some aspect of college life that interests you. Consider using the online library of a college you are evaluating or an article from the Prentice Hall Supersite. Print a copy of that article, and use a highlighter and the PQ3R method to read it.

Endnotes

1 Thomas M. Smith, U.S. Department of Education, National Center for Education Statistics, *The Condition of Education 1996,* NCES 96-304 (Washington, DC: U.S. Government Printing Office, 1996), 84.

2 Sherwood Harris, *The New York Public Library Book of How and Where to Look It Up* (Englewood Cliffs, NJ: Prentice Hall, 1991), 13.

3 George M. Usova, *Efficient Study Strategies: Skills for Successful Learning* (Pacific Grove, CA: Brooks/Cole Publishing Company, 1989), 45.

4 Francis P. Robinson, *Effective Behavior* (New York: Harper & Row, 1941).

5 Sylvan Barnet and Hugo Bedau, *Critical Thinking, Reading, and Writing: A Brief Guide to Argument,* 2nd ed. (Boston: Bedford Books of St. Martin's Press, 1996), 15–21.

6 John J. Macionis, *Sociology,* 6th ed. (Upper Saddle River, NJ: Prentice-Hall, 1997). Reprinted by permission of Prentice-Hall, Inc.

7 T. Audesirk and G. Audesirk, *Life on Earth* (Upper Saddle River, NJ: Prentice-Hall, 1997). Reprinted by permission of Prentice-Hall, Inc.

F O C U S

8

Listening and Note Taking

College exposes you daily to facts, opinions, and ideas. It is up to you to take in, retain, and demonstrate your knowledge of what you learn. Two important skills—listening and note taking—will help you internalize and remember valuable information you need. In this chapter, you will learn strategies to improve your ability to take in information through active listening and record information through note taking.

In this chapter, you will explore answers to the following questions:

- Why is listening a skill?
- What are the stages of listening?
- How can I improve my listening skills?
- How does note taking help me?
- What are the three steps to effective note taking?
- What note-taking system should I use?
- How can I write faster when taking notes?

WHY IS LISTENING A SKILL?

The act of hearing isn't quite the same as the act of listening. Hearing refers to sensing spoken messages from their source; listening involves a complex process of communication. Successful listening results in the speaker's intended message reaching the listener. In school and at home, poor listening results in communication breakdowns and mistakes, and skilled listening promotes progress and success.

Ralph G. Nichols, a pioneer in listening research, studied two hundred students at the University of Minnesota over a nine-month period. His findings, summarized in Table 8.1, demonstrate that effective listening depends as much on a positive attitude as on specific skills.[1]

Listening is a teachable—and learnable—skill. Improving your learning skills involves understanding the stages of listening, managing listening challenges, and becoming an active listener. Although becoming a better listener will help in every class, it is especially important in subject areas that are difficult for you.

TABLE 8.1 What helps and hinders listening.

LISTENING IS HELPED BY . . .	LISTENING IS HINDERED BY . . .
making a conscious decision to work at listening; viewing difficult material as a listening challenge	caring little about the listening process; tuning out difficult material
fighting distractions through intense concentration	refusing to listen at the first distraction
continuing to listen when a subject is difficult or dry, in the hope that one might learn something interesting	giving up as soon as one loses interest
withholding judgment until hearing everything	becoming preoccupied with a response as soon as a speaker makes a controversial statement
focusing on the speaker's theme by recognizing organizational patterns, transitional language, and summary statements	getting sidetracked by unimportant details
adapting note-taking style to the unique style and organization of the speaker	always taking notes in outline form, even when a speaker is poorly organized, leading to frustration
pushing past negative emotional responses and forcing oneself to continue to listen	letting an initial emotional response shut off continued listening
using excess thinking time to evaluate, summarize, and question what one just heard and anticipating what will come next	thinking about other things and, as a result, missing much of the message

To overcome barriers, explore what the listening process is and the reasons it can be hard to listen well.

WHAT ARE THE STAGES OF LISTENING?

Listening is made up of four stages that build on one another: sensing, interpreting, evaluating, and reacting. These stages take the message from the speaker to the listener and back to the speaker (see Figure 8.1):

- During the sensation stage (also known as hearing), your ears pick up sound waves and transmit them to the brain. For example, you are sitting in class and hear your instructor say, "The only opportunity to make up last week's test is Tuesday at 5 P.M."
- In the interpretation stage, listeners attach meaning to a message. This involves understanding what is being said and relating it to what you already know. For example, when you hear this message, you relate it to your knowledge of the test, whether you need to make it up, and what you are doing on Tuesday at 5 P.M.
- In the evaluation stage of listening, you decide how you feel about the message—whether, for example, you like it or agree with it. This involves considering the message as it relates to your needs and values. In this example, if you do need to make up the test but have to work Tuesday at 5 P.M., you evaluate that you aren't thrilled with the message.
- The final stage of listening involves a reaction to the message in the form of direct feedback. Your reaction, in this example, may be to raise your hand or stick around after class and ask the instructor if there is any alternative to that particular time for the makeup test.

Stages of listening. FIGURE 8.1

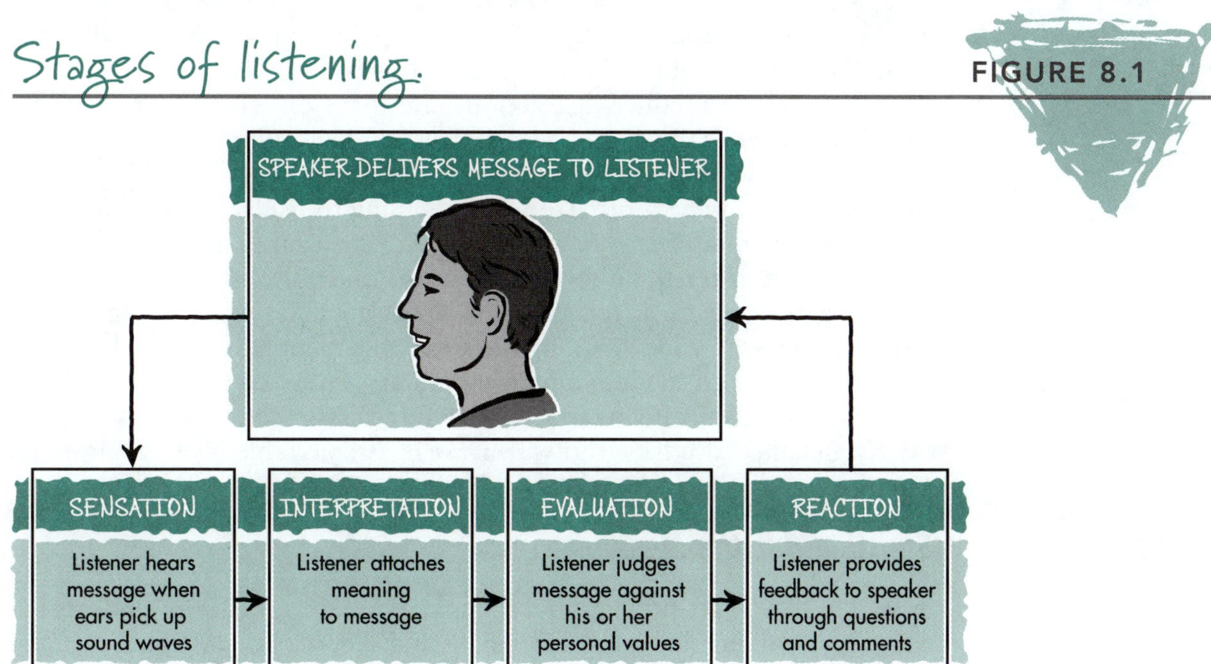

HOW CAN I IMPROVE MY LISTENING SKILLS?

Mastering your listening skills has many advantages, including improving comprehension during lectures and making note taking easier. Improving your listening skills involves managing listening challenges and becoming an active listener.

Manage Listening Challenges

Classic studies have shown that immediately after listening, students are likely to recall only half of what was said. This is partly due to particular listening challenges, including divided attention and distractions, the tendency to shut out the message, the inclination to rush to judgment, and partial hearing loss or learning disabilities.[2] To help create a positive listening environment in both your mind and your surroundings, explore how to manage these challenges.

Divided Attention and Distractions

Imagine you are talking with a friend at a noisy party when, suddenly, you hear your name mentioned across the room. You weren't consciously listening to anything outside your own conversation. Now, though, you strain to hear what someone might be saying about you and listen with only half an ear to what your friend says. Chances are, you hear neither person very well. This situation illustrates the consequences of divided attention. Although you are capable of listening to more than one message at the same time, the usual result is that you may not completely hear or understand any of them.

Internal and external distractions often divide your attention. Internal distractions include anything from hunger to headache to personal worries. Something the speaker says may also trigger a recollection that may cause your mind to drift. In contrast, external distractions include noises (whispering, honking horns, screaming sirens) and even excessive heat or cold. It can be hard to listen in an overheated room that is putting you to sleep.

Your goal is to reduce distractions and focus on what you're hearing. Sitting where you can see and hear clearly will help. When you can see and hear your instructors, you have a much better chance of being able to listen well, and you may be more likely to be willing to listen. In order to avoid activity that might divide your attention, you may want to sit away from people who might distract you by chatting or making noise.

Make sure you are as relaxed and alert as possible. Work to concentrate on class when you're in class, and save worrying about personal problems for later. Try not to go to class hungry or thirsty. Dress comfortably. Bring a sweater or sweatshirt if you anticipate that the classroom will be too cold. If there's a chance you'll be too warm, wear a removable layer of clothing.

Shutting Out the Message

Instead of paying attention to everything the speaker says, many students fall into the trap of focusing on specific points and shutting out the rest of the message. Worse, if you perceive that a subject is too difficult or uninteresting, you may tune out everything. Shutting out the message makes it tough to listen well from that point on because the information you miss may be the foundation for what goes on in future classes. Creating a positive listening environment includes accepting responsibility for listening. The instructor communicates information to you, but he or she cannot force you to listen. You are responsible for taking in that information. One important motivator is believing that what your instructors say is valuable. As many students learn the hard way, instructors often cover material from outside the textbook during class and then test on that material. If you work to take in the whole message in class, you will be able to read over your notes later and think critically about what is most important.

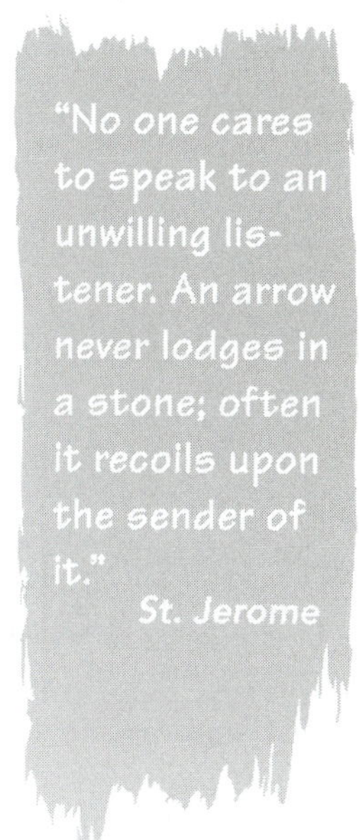

The Rush to Judgment

People tend to stop listening when they hear something they don't like. If you rush to judge what you've heard, your focus turns to your personal reaction rather than the content of the speaker's message. Students who disagree during a lecture often spend a lot of thinking time figuring out how to word a question or comment for response.

Judgments also involve reactions to the speakers themselves. If you do not like your instructors or if you have preconceived notions about their ideas or cultural background, you may decide that their words have little value. Anyone whose words have ever been ignored because of race, background, gender, or disability understands how prejudice can interfere with communication.

Work to recognize and control your judgments. Being aware of what you tend to judge will help you avoid putting up a barrier against incoming messages that clash with your opinions or feelings. Keeping an open mind means being aware of the things you believe in as well as your prejudices. It also means defining education as a continuing search for evidence, regardless of whether it supports or negates your point of view.

Partial Hearing Loss and Learning Disabilities

Good listening techniques don't solve every listening problem. Students who have a partial hearing loss have a physical explanation for why listening is difficult. If you have some level of hearing loss, seek out special services that can help you listen in class. You may require special equipment or might benefit from tutoring. You may be able to arrange to meet with your instructor outside of class to clarify your notes.

Other disabilities, such as attention deficit disorder (ADD) or a problem with processing heard language, can cause difficulties with both focusing on and understanding that which is heard. People with such disabilities have varied ability to compensate for and overcome them. If you have a disability, don't blame yourself for having trouble listening. When you get to college, your counseling center, student health center, advisor, and instructors will be able to give you particular assistance in working through your challenges.

Become an Active Listener

On the surface, listening seems like a passive activity: You sit back and listen as someone else speaks. Effective listening, however, is really an active process that involves setting a purpose for listening, asking questions, and paying attention to verbal signposts.

Set Purposes for Listening

Active listening is difficult if you don't know or care why you are listening. Think through why you listen in any situation. Establish what you want to achieve by listening, such as greater understanding of the material, a more direct connection with your instructor, better alertness in class, or better note taking. When you set a purpose, you have a goal that you can achieve only through active listening. A purpose for listening motivates you to listen.

Ask Questions

Asking questions is not a sign of stupidity or a reason to doubt your intelligence. In fact, a willingness to ask questions shows a desire to learn and is the mark of an active listener and critical thinker. Some questions are informational—seeking information—such as any question beginning with the phrase, "I don't understand . . . " Other clarifying questions state your understanding of what you just heard and ask if that understanding is correct. Some clarifying questions focus on a key concept or theme ("So, some learning disorders can be improved with treatment?"); others highlight specific facts ("Is it true that dyslexia can cause people to reverse letters and words?").

If, for whatever reason, you don't have an opportunity to ask your questions in class, jot them down and ask them during a discussion period or during a talk with your instructor.

Pay Attention to Verbal Signposts

You can identify important facts and ideas and predict test questions by paying attention to the speaker's specific choice of words. Verbal signposts often involve transition words and phrases that help organize information, connect ideas, and indicate what is important and what is not. Let phrases like those in Table 8.2 direct your attention to the material that follows.

Paying attention to verbal signposts.

SIGNALS POINTING TO KEY CONCEPTS	SIGNALS OF SUPPORT
There are two reasons for this . . .	For example, . . .
A critical point in the process involves . . .	Specifically, . . .
Most importantly, . . .	For instance, . . .
The result is . . .	Similarly, . . .
SIGNALS POINTING TO DIFFERENCES	**SIGNALS OF SUMMARIES**
On the contrary, . . .	Finally, . . .
On the other hand, . . .	Recapping this idea, . . .
In contrast, . . .	In conclusion, . . .
However, . . .	As a result, . . .

Source: Adapted from George M. Usova, *Efficient Study Strategies: Skills for Successful Learning* (Pacific Grove, CA: Brooks/Cole Publishing Company, 1989), p. 69.

Effective listening will enable you to acquire knowledge and assist you in note taking. As you sit through classroom lectures, actively use the listening strategies to get the most from your experience.

HOW DOES NOTE TAKING HELP ME?

Notes help you learn when you are listening in class, doing research, or studying. Because it is virtually impossible to take notes on everything you hear or read, the act of note taking encourages you to decide what is worth remembering. The positive effects of note taking include the following:

- Your notes provide material that helps you study information and prepare for tests.
- When you take notes, you become an active, involved listener and learner.
- Notes help you think critically and organize ideas.
- The information you learn in class may not appear in any text; you will have no way to study it without writing it down.
- If it is difficult for you to process information while in class, having notes to read and make sense of later can help you learn.
- Note taking is a skill for life that you will use on the job and in your personal life.

With so many advantages to taking good notes, it is imperative that you learn effective methods of note taking.

WHAT ARE THE THREE STEPS TO EFFECTIVE NOTE TAKING?

Your notes have two purposes: First, they should reflect what you heard in class; second, they should be a resource for studying, writing, or comparing with your text material. Because it is virtually impossible to take notes on everything you hear or read, the act of note taking encourages you to decide what is worth remembering. To make the most effective use of your notes, you should follow three steps: preparing to take notes prior to class, recording information in class, and reviewing information after class.

Preparing to Take Notes

Taking good class notes depends on good preparation. The following is a guide to good preparation:

- Complete the reading before class so that the lecture becomes more of a review than an introduction.
- While you are reading, mark the text, using highlighters, and write notes and questions in the margin of your text.
- Use separate pieces of 8 1/2 by 11-inch paper for each class. If you use a three-ring binder, punch holes in papers your instructor hands out, and insert them immediately following your notes for that day.

- Take a comfortable seat where you can easily see and hear, and be ready to write as soon as the instructor begins speaking.
- Choose a note-taking system that helps you handle the instructor's speaking style. One instructor may deliver organized lectures at a normal speaking rate; another may jump from topic to topic or talk very quickly.
- Set up a support system with a student in each class. That way, if you are absent, you can get the notes you missed.

Following these simple steps will prepare you to get the most important information recorded.

Recording Information in Class

Because no one has the time to write down everything he or she hears, the following strategies will help you choose and record what you feel is important, in a format that you can read and understand later.

- Date each page. When you take several pages of notes during a lecture, add an identifying letter or number to the date on each page: 11/27 A, 11/27 B, . . . or 11/27—1 of 3, 11/27—2 of 3.
- Add the specific topic of the lecture at the top of the page. For example:
  11/27A—*U.S. Immigration Policy After World War II.*
- If your instructor jumps from topic to topic during a single class, try starting a new page for each new topic.
- Ask yourself critical-thinking questions as you listen: Do I need this information? Is the information important, or is it just a digression? Is the information fact or opinion? If it is opinion, is it worth remembering?
- Record whatever an instructor emphasizes (see Figure 8.2 for details).

FIGURE 8.2 How to pick up on instructor cues.

- Continue to take notes during class discussions and question-and-answer periods. What your fellow students ask about may help you as well.
- Leave one or more blank spaces between points. This white space will help you review your notes because information will appear in self-contained sections.
- Draw pictures and diagrams that help illustrate ideas.
- Indicate material that is especially important by using a star, underlining, using a highlighter pen or writing words in capital letters.
- If you cannot understand what the instructor is saying, leave a space and place a question mark in the margin. Then ask the instructor to explain it again after class, or discuss it with a classmate. Fill in the blank when the idea is clear.
- Take notes until the instructor stops speaking. Students who stop writing a few minutes before the class is over can miss critical information.
- Make your notes as legible, organized, and complete as possible. Your notes are only useful if you can read and understand them.

"Omit needless words. . . .This requires not that the writer make all his sentences short, or that he avoid all detail and treat his subjects only in outline, but that every word tell."
William Strunk, Jr.

Even though recording information is an important and necessary step in note taking, reviewing your notes becomes as valuable.

Reviewing Notes After Class

Class notes are a valuable study tool when you review them regularly. Effective review begins within a day of the lecture. Studies show that it is much more effective to review for 10 to 15 minutes each day rather than attempt to review for long periods of time the night before a test. Follow these steps:

- Read over the notes to learn the information, clarify abbreviations, fill in missing information, and underline or highlight key points.
- Summarize your notes by critically evaluating which ideas and examples are most important and by writing them in a condensed form.
- Recite the information in your notes, and explain the concepts in your own words.

You can take notes in many ways. Different note-taking systems suit different people and situations. Explore each system and choose what works for you.

WHAT NOTE-TAKING SYSTEM SHOULD I USE?

You will benefit most from the system that feels most comfortable to you. The most common note-taking systems include outlines, the Cornell system, and think links.

Taking Notes in Outline Form

When a reading assignment or lecture seems well organized, you may choose to take notes in outline form. Outlining shows the relationships among ideas and their supporting examples through the use of line-by-line phrases set off by varying indentations.

FIGURE 8.3 Sample formal outline.

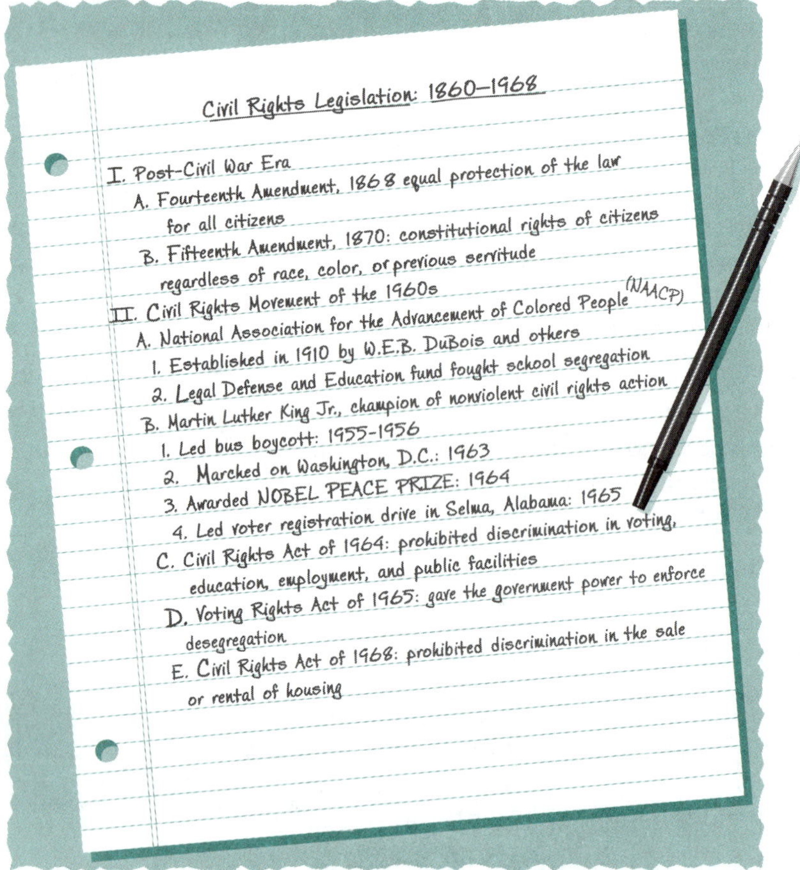

Formal outlines indicate ideas and examples using Roman numerals, capital and lowercase letters, and numbers. Formal outlines also require at least two headings on the same level—that is, if you have a II A, you must also have a II B. Figure 8.3 shows an outline on civil rights legislation. Many times, however, lectures don't follow a formal outline, and a listener can become caught up in trying to create a perfect outline rather than listening to and recording information. When lectures don't follow a formal outline or when you are pressed for time, such as during class, you can use an informal system of consistent indenting and dashes instead.

Using the Cornell Note-Taking System

The Cornell note-taking system, also known as the T-note system, was developed more than forty-five years ago by Walter Pauk at Cornell University.[3] The system is successful because it is simple—and because it works. It consists of three sections on ordinary note paper:

- Section 1, the largest section, is on the right. Record your notes here in informal outline form.
- Section 2, to the left of your notes, is the cue column. Leave it blank while you read or listen, then fill it in later as you review. You might fill it

with comments that highlight main ideas, clarify meaning, suggest examples, or link ideas and examples. You can even draw diagrams.

- Section 3, at the bottom of the page, is the summary area, where you summarize the notes on the page. When you review, use this section to reinforce concepts and provide an overview. Make an upside-down letter T and use Figure 8.4 as your guide. Make the cue column about 2 1/2 inches wide and the summary area 2 inches tall. Figure 8.4 shows how a student used the Cornell system to take notes in an Introduction to Business course.

When you use the Cornell system, create the note-taking structure before class begins.

Notes taken using the Cornell system. FIGURE 8.4

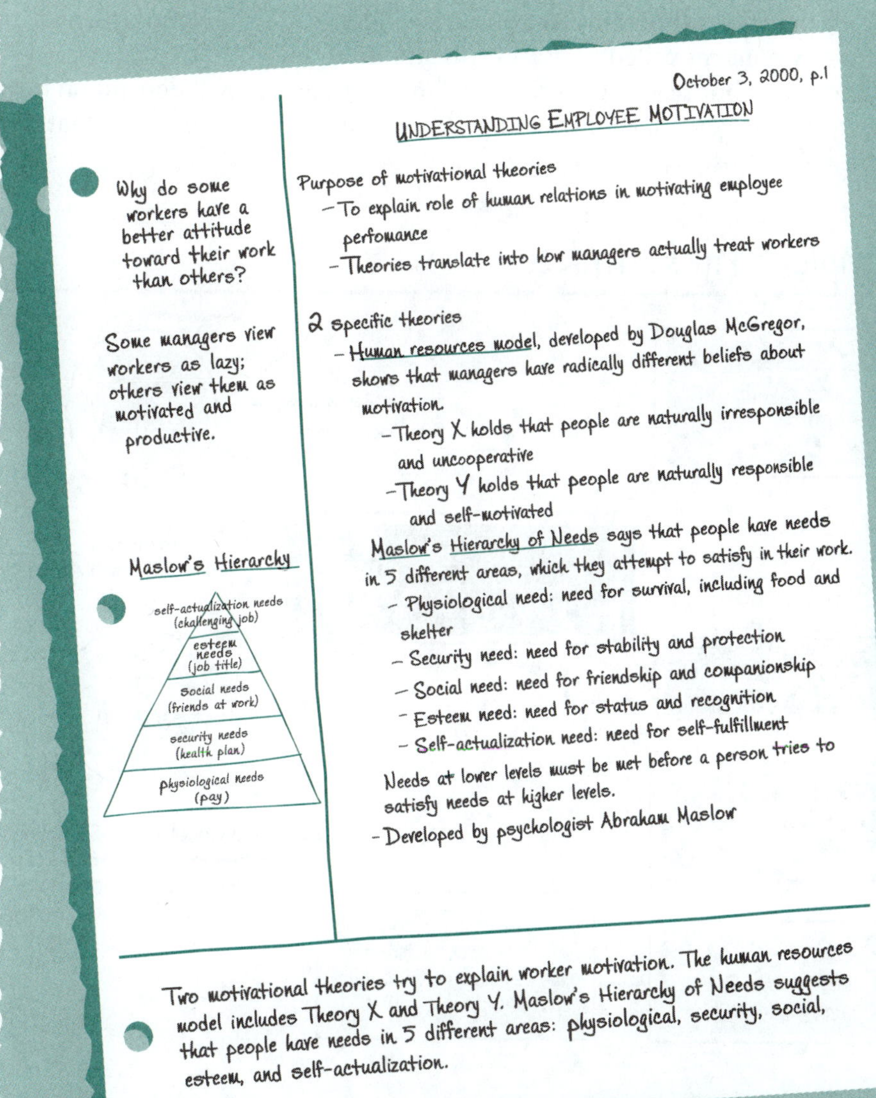

October 3, 2000, p.1

UNDERSTANDING EMPLOYEE MOTIVATION

Why do some workers have a better attitude toward their work than others?

Purpose of motivational theories
- To explain role of human relations in motivating employee perfomance
- Theories translate into how managers actually treat workers

Some managers view workers as lazy; others view them as motivated and productive.

2 specific theories
- Human resources model, developed by Douglas McGregor, shows that managers have radically different beliefs about motivation.
  - Theory X holds that people are naturally irresponsible and uncooperative
  - Theory Y holds that people are naturally responsible and self-motivated

Maslow's Hierarchy

Maslow's Hierarchy of Needs says that people have needs in 5 different areas, which they attempt to satisfy in their work.
- Physiological need: need for survival, including food and shelter
- Security need: need for stability and protection
- Social need: need for friendship and companionship
- Esteem need: need for status and recognition
- Self-actualization need: need for self-fulfillment

Needs at lower levels must be met before a person tries to satisfy needs at higher levels.
- Developed by psychologist Abraham Maslow

Two motivational theories try to explain worker motivation. The human resources model includes Theory X and Theory Y. Maslow's Hierarchy of Needs suggests that people have needs in 5 different areas: physiological, security, social, esteem, and self-actualization.

Creating a Think Link

A think link (also known as a mind map, a web, or a cluster) is a visual form of note taking. When you draw a think link, you diagram ideas using shapes and lines that link ideas and supporting details and examples. The visual design makes the connections easy to see, and the use of shapes and pictures extends the material beyond just words. Many learners respond well to the power of visualization. You can use think links to brainstorm ideas for paper topics as well.

One way to create a think link is to start by circling your topic in the middle of a sheet of unlined paper. Next, draw a line from the circled topic, and write the name of the first major idea at the end of that line. Circle the idea also. Then jot down specific facts related to the idea, linking them to the idea with lines. Continue the process, connecting thoughts to one another using circles, lines, and words.

A think link may be difficult to construct in class, especially if your instructor talks quickly. In this case, use another note-taking system during class. Then make a think link as you review. Figure 8.5 shows a think link on a sociology concept called social stratification.

Once you choose a note-taking system, your success will depend on how well you use it. Personal shorthand will help you make the most of whatever system you choose.

FIGURE 8.5 Sample think link.

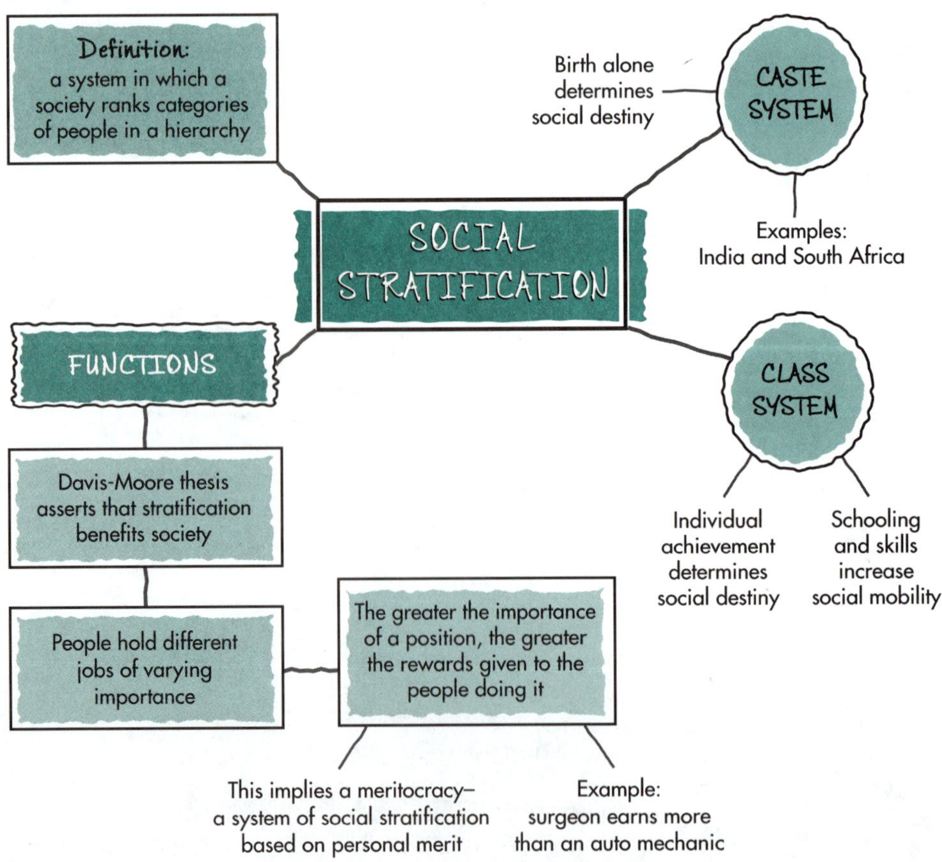

HOW CAN I WRITE FASTER WHEN TAKING NOTES?

When taking notes, many students feel they can't keep up with the instructor. Using some personal shorthand (not standard secretarial shorthand) can help to push the pen faster. Shorthand is writing that shortens words or replaces them with symbols. Because you are the only intended reader, you can misspell and abbreviate words in ways that only you understand.

The only danger with shorthand is that you might forget what your writing means. To avoid this problem, review your shorthand notes while your abbreviations and symbols are fresh in your mind. If there is any confusion, spell out words as you review.

Here are six suggestions that will help you master this important skill:

1. Use the following standard abbreviations in place of complete words:

w/	with	cf	compare; in comparison to
w/o	without	ff	following
→	means; resulting in	Q	question
←	as a result of	p.	page
↑	increasing	*	most importantly
↓	decreasing	<	less than
∴	therefore	>	more than
∵	because	=	equals
≈	approximately	%	percent
+ or &	and	Δ	change
—	minus; negative	2	to; two; too
No. or #	number	vs.	versus; against
i.e.	that is	e.g.	for example
etc.	and so forth	c/o	in care of
ng	no good	lb	pound

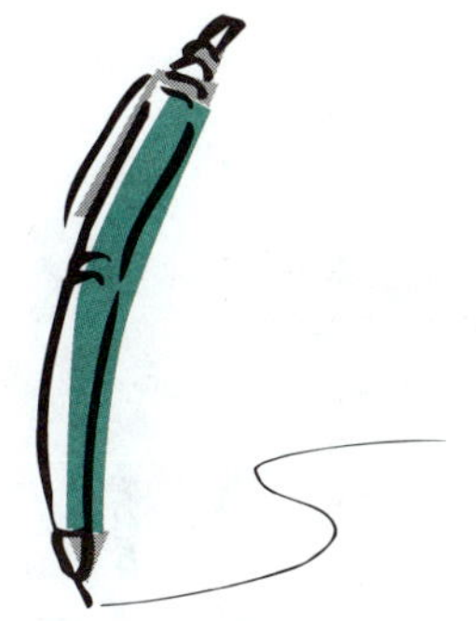

2. Shorten words by removing vowels from the middle of words:

   prps = purpose

   Crvtte = Corvette (as on a vanity license plate for a car)

3. Form plurals by adding s:

   prblms = problems   prntrs = printers

4. Substitute word beginnings for entire words:

   assoc = associate; association

   info = information

5. Make up your own symbols and use them consistently:

   b/4 = before   2thake = toothache

6. Use key phrases instead of complete sentences ("German—nouns capitalized" instead of "In the German language, all nouns are capitalized").

Note taking is a valuable tool to learn before you enter college. As you take time to learn this skill, experiment with systems and methods that work best for you. Having good notes is one of the many keys to success.

EXERCISES AND ACTIVITIES

8.1 Evaluating My Notes

Choose one particular class period from the last two weeks. Have a classmate photocopy his or her notes from that class period for you. Then evaluate your notes by comparing them with your classmate's. Ask yourself:

- Do my notes make sense?
- How is my handwriting?
- Do the notes cover everything that was brought up in class?
- Are there examples to back up ideas?
- What note-taking system is used?
- Will these notes help me study?

Write your evaluation on a separate piece of paper.

What ideas or techniques from your classmate's notes do you plan to use in the future?

8.2 Class Versus Reading

Pick a class for which you have a regular textbook. Choose a set of class notes on a subject that is also covered in that textbook. Read the textbook section that corresponds to the subject of your class notes, taking notes as you go. Compare your reading notes to the notes you took in class.

Did you use a different system with the textbook or the same as in class? Why?

Which notes can you understand better? Why do you think that's true?

What did you learn from your reading notes that you want to bring to your class note-taking strategy?

8.3 Take Notes on the News

Videotape a news program and take notes while you view it. Review the tape and compare it to your notes. How accurate and complete were your notes? Repeat the process using each of the suggested note-taking methods.

Endnotes

1 Ralph G. Nichols, "Do We Know How to Listen? Practical Helps in a Modern Age," *Speech Teacher* (March 1961): 118–124.

2 Ibid.

3 Walter Pauk, *How to Study in College,* 5th ed. (Boston: Houghton Mifflin Company, 1993), 110–114.

E X C E L

E X C E L

9

Test Taking

Tests might be considered an unpleasant but necessary task of everyone's high school and college experience. It is rare that you have a class in which no tests are given. Tests are used to assess how well you have learned the information in a given time period. Although tests can be stressful, there are specific strategies you can learn that will improve your test-taking ability.

In this chapter, you will explore answers to the following questions:

- What kind of preparation helps improve test scores?
- What strategies can help me succeed on tests?
- How can I learn from test mistakes?

WHAT KIND OF PREPARATION HELPS IMPROVE TEST SCORES?

Many people don't look forward to taking tests. If you are one of those people, try thinking of exams as preparation for life. When you volunteer, get a job, or work on your personal budget, you'll have to apply what you know. This is exactly what you do when you take a test.

Like a runner who prepares for a marathon by exercising, eating right, taking practice runs, and getting enough sleep, you can take steps to master your exams. Your first step is to study until you know the material that will be on the test. Your next step is to use the following strategies to become a successful test taker: Identify test type, use specific study skills, prepare physically, and conquer test anxiety.

Identify Test Type and Material Covered

Before you begin studying, try to determine the type of test and what it will cover:

- Will it be a short-answer test with true/false and multiple-choice questions, an essay test, or a combination?
- Will the test cover everything you studied since the semester began, or will it be limited to a narrow topic?
- Will the test be limited to what you learned in class and in the text, or will it also cover outside readings?

Your instructors may answer these questions for you. Even though they may not tell you the specific questions that will be on the test, they might let you know what blocks of information will be covered and the question format. When instructors write information on the board, that is a pretty clear signal that they feel the information is important. Some instructors may even drop hints throughout the semester about possible test questions. Some comments are direct ("I might ask a question on the subject of __________ on your next exam"); other clues are subtle. For example, when instructors repeat an idea or when they express personal interest in a topic ("One of my favorite theories is . . . "), they are letting you know that the material may be on the test.

Here are a few other strategies for predicting what may be on a test:

- Use PQ3R to identify important ideas and facts. Often, the questions you write and ask yourself when you read assigned materials may be part of the test. In addition, any textbook study questions are good candidates for test material.
- If you know people who took the instructor's course before, ask them about class tests. Try to find out how difficult the tests are and whether the test focuses more on assigned readings or class notes. Ask about instructor preferences. If you learn that the instructor pays close attention to detail such as facts or grammar, plan your work accordingly.
- Examine previous tests if instructors make them available in class or on reserve in the library. If you can't get copies of old tests, use clues from the

class to predict test questions. After taking the first exam in the course, you will have a lot more information about what to expect in the future.

Use Specific Study Skills

Certain study skills are especially useful for test taking. They include choosing study materials, setting a study schedule, critical thinking, taking a pretest, and becoming organized.

Choose Study Materials

Once you have identified as much as you can about the subject matter of the test, choose the materials that contain the information you need to study. You can save yourself time by making sure that you aren't studying anything you don't need to. Go through your notes, your texts, any primary source materials that were assigned, and any handouts from your instructor. Set aside any materials you don't need so they don't take up your valuable time.

Set a Study Schedule

Use time management skills discussed in Chapter 8 to set a schedule that will help you feel as prepared as you can be. Consider all the relevant factors—the materials you need to study, how many days or weeks until the test date, and how much time you can study each day. If you establish your schedule ahead of time and write it in your date book, you will be much more likely to follow it.

Schedules will vary widely according to situation. For example, if you have only three days before the test and no other obligations during that time, you might set a two-hour study session for yourself during each day. On the other hand, if you have two weeks before a test date, classes during the day, and work three nights a week, you might spread out your study sessions over the nights you have off work during those two weeks. If you truly want to learn the material, don't cram! Stick to your schedule, so your review time will be productive.

Prepare Through Critical Thinking

Approach your test preparation critically, working to understand rather than just to pass the test by repeating facts. As you study, try to connect ideas to examples, analyze causes and effects, establish truth, and look at issues from different perspectives. Although it takes work, critical thinking will promote a greater understanding of the subject and probably a higher grade on the exam. Using critical thinking is especially important for essay tests. Prepare by identifying potential essay questions, planning your answers, and writing your responses.

Take a Pretest

Use questions from the ends of textbook chapters to create your own pretest. Choose questions that are likely to be covered on the test; then answer them under testlike conditions—in quiet, with no books or notes to help you, and with a clock telling you when to quit. Try to duplicate the conditions of the actual test. If your course doesn't have an assigned text, develop questions from your notes and from assigned outside readings.

Become Organized

A checklist, like the one in Figure 9.1, will help you get organized and stay on track as you prepare for each test.

Prepare Physically

When taking a test, you often need to work efficiently under time pressure. If your body is tired or under stress, you will probably not think as clearly or perform as well. If you can, avoid pulling an all-nighter. Get some sleep so that you can wake up rested and alert. If you are one of the many who press the snooze button in their sleep, try setting two alarm clocks and placing them across the room from your bed. That way, you'll be more likely to get to school and to your test on time.

Eating right is also important. Sugar-laden snacks will bring your energy up only to send it crashing back down much too soon. Similarly, too much caffeine can add to your tension and make it difficult to focus. Eating nothing will leave you drained, but too much food can make you want to take a nap. The best advice is to eat a light, well-balanced meal before a test. When time is short, grab a quick-energy snack such as a banana, some orange juice, or a granola bar.

Conquer Test Anxiety

A certain amount of stress can be a good thing. Your body is on alert, and your energy motivates you to do your best. For many students, however, the time before and during an exam brings a feeling of near-panic known as test anxiety. Described as a bad case of nerves that makes it hard to think or remember, test anxiety can make your life miserable and affect how you perform on tests. When anxiety blocks performance, here are some suggestions:

- Prepare so you'll feel in control. The more you know about what to expect on the exam, the better you'll feel. Find out what material will be covered, the format of the questions, the length of the exam, and the percentage of points assigned to each question.
- Put the test in perspective. No matter how important it may seem, a test is only a small part of your educational experience and an even smaller part of your life. Your test grade does not reflect the kind of person you are or your ability to succeed in life.
- Make a study plan. Divide the plan into a series of small tasks. As you finish each one, you'll feel a sense of accomplishment and control.
- Practice relaxation. When you feel test anxiety coming on, take some deep breaths, close your eyes, and visualize positive mental images related to the test, like getting a good grade and finishing confidently with time to spare.

Pretest checklist.

FIGURE 9.1

Course: ______________________ Teacher: ________________

Date, time, and place of test: ______________________________

Type of test (e.g., Is it a midterm or a minor quiz?): ____________

What the instructor has told you about the test, including the types of test questions, the length of the test, and how much the test counts in your final grade: ______________________________

Topics to be covered on the test in order of importance:

1. ______________________________
2. ______________________________
3. ______________________________
4. ______________________________
5. ______________________________

Study schedule, including materials you plan to study (e.g., texts and class notes) and date you plan to complete each source:

Source:	Date of Completion:
1.	
2.	
3.	
4.	
5.	

Materials you are expeced to bring to the test (e.g., your textbook, a sourcebook, a calculator): ______________________________

Special study arrangements (e.g., plan study group meetings, ask the instructor for special help, get outside tutoring): ____________

Life management issues (e.g., rearrange work hours): ____________

Coping with Math Anxiety

For many students, there is a special anxiety associated with taking a math test. As Sheila Tobias, author of *Overcoming Math Anxiety,* explains, math anxiety is linked to the feeling that math is impossible:

> The first thing people remember about failing at math is that it felt like sudden death. Whether it happened while learning word problems in sixth grade, coping with equations in high school, or first confronting calculus and statistics in college, failure was instant and frightening. An idea or a new operation was not just difficult, it was impossible! And instead of asking questions or taking the lesson slowly, assuming that in a month or so they would be able to digest it, people remember the feeling, as certain as it was sudden, that they would never go any further in mathematics.[1]

Students who believe they are no good at math probably won't do well on math tests, even if they study. Their attitude creates a huge problem. If you are one of these students, here are some steps you can take to begin thinking about math—and math tests—in a different way:

- See the value in learning to use your mind in a mathematical way. Mathematical thinking is another type of critical thinking. It can help you solve the little and big problems that are or will be part of your world, including how to measure the amount of wallpaper you need in a room, how to compare student loan programs, and even how to analyze the stock market.
- Think of math as a tool that will help you land a good job. In fields such as engineering, accounting, banking, and the stock market, the ability to solve numerical problems is at the heart of the work. In real estate, retail sales, medicine, and publishing, you may use math for tasks such as writing budgets and figuring mortgage rates.
- Turn negative self-talk into positive self-talk. Instead of telling yourself that a problem is too hard, tell yourself that if you take small, logical steps, you will succeed. Says Tobias, "If we can talk ourselves into feeling comfortable and secure, we may let in a good idea."[2]
- Don't believe that women can't do math. Sheila Tobias says that when male students fail a math quiz, they think they didn't work hard enough, but when female students fail, they are three times more likely to feel that they just don't have what it takes.[3] Whether you are a male or a female, work to overcome this stereotype.
- Use the people and resources around you. Get to know your math instructor so you're comfortable asking for help. Join a math study group, and make building confidence a group goal. Have a pep meeting right before a big test. Seek out a classmate or a tutor who can help you with your skills and build your confidence.
- Become comfortable in the world of math. Find a computer program with math games, or buy a paperback book with math puzzles. Do percentages and estimations in your head. Have fun with problems and enjoy solving them. Then transfer these feelings to class work and tests.
- Understand math's relationship to your life success. Being at ease with numbers can serve you in day-to-day functions. Percentages can help you compare the financial benefits of different loan programs; adding

and subtracting will allow you to balance a checkbook, and fractions will help you compare costs at work. Furthermore, working with numbers helps to develop general thinking skills. The precise calculation and problem solving involved in math help you develop precision, a focus on detail, patience, and a sense of order.

When you have prepared using the strategies that work for you, you are ready to take your exam. Focus on methods that can help you succeed when the test begins.

WHAT STRATEGIES CAN HELP ME SUCCEED ON TESTS?

Even though every test is different, there are general strategies that will help you handle almost all tests, including short-answer and essay exams.

Write Down Key Facts

Before you even look at the test, write down any key information—including formulas, rules, and definitions—that you studied recently or even right before you entered the classroom for the test. Use the back of the question sheet or a piece of scrap paper for your notes. (Make sure it is clear to your instructor that this scrap paper didn't come into the classroom already filled in!) Recording this information right at the start will make forgetting less likely.

Begin with an Overview of the Exam

Even though exam time is precious, spend a few minutes at the start of the test to get a sense of the kinds of questions you'll be answering, what kind of thinking they require, the number of questions in each section, and the point value of each section. Use this information to schedule the time you spend on each section. For example, if a two-hour test is divided into two sections of equal point value—an essay section with four questions and a short-answer section with sixty questions—you can spend an hour on the essays (fifteen minutes per question) and an hour on the short-answer section (one minute per question).

As you make your calculations, think about the level of difficulty of each section. If you think you can handle the short-answer questions in less than an hour and that you'll need more time with the essays, rebudget your time in a way that works for you.

Know the Ground Rules

A few basic rules apply to any test, and following them will give you the advantage:

- Read test directions. Although a test made up of one hundred true/false questions and one essay may look straightforward, the directions may tell you to answer eighty or that the essay is an optional bonus. Some questions or sections may be weighted more heavily than others. Try circling or underlining key words and numbers that remind you of the directions.

- Begin with the parts or questions that seem easiest to you. Starting with what you know best can boost your confidence and help you save time to spend on the harder parts.
- Watch the clock. Keep track of how much time is left and how you are progressing. You may want to plan your time on a scrap piece of paper, especially if you have one or more essays to write. Wear a watch or bring a small clock with you to the classroom. A wall clock may be broken, or there may be no clock at all! Also, take your time. Rushing is almost always a mistake, even if you feel you've done well. Stay till the end so you can refine and check your work.
- Master the art of intelligent guessing. When you are unsure of an answer, you can leave it blank, or you can guess. In most cases, guessing will benefit you. First, eliminate all the answers you know—or believe—are wrong. Try to narrow your choices to two possible answers; then choose the one that makes more sense to you. When you recheck your work, decide if you would make the same guesses again, making sure there isn't a qualifier or fact that you hadn't noticed before.
- Follow directions on machine-scored tests. Machine-scored tests require that you use a special pencil to fill in a small box on a computerized answer sheet. Use the right pencil (usually a number 2), and mark your answer in the correct space. Neatness counts on these tests because the computer can misread stray pencil marks or partially erased answers. Periodically, check the answer number against the question number to make sure they match. One question skipped can cause every answer following it to be marked incorrect.

The general strategies you have just explored also can help you address specific types of test questions.

Master Different Types of Test Questions

Objective questions
Short-answer questions that test your ability to recall, compare, and contrast information and to choose the right answer from a limited number of choices.

Although the goal of all test questions is to discover how much you know about a subject, every question type has its own way of asking what you know. **Objective questions,** such as multiple-choice or true/false, test your ability to recall, compare, and contrast information and to choose the right answer from among several choices. **Subjective questions,** usually essay questions, demand the same information recall but ask that you use critical-thinking strategies to answer the question, then organize, draft, and refine a written response. The following guidelines will help you choose the best answers to both types of questions.

Multiple-Choice Questions

Subjective questions
Essay questions that require you to express your answer in terms of your own personal knowledge and perspective.

Multiple-choice questions are the most popular type on standardized tests. The following strategies can help you answer these questions:

- Read the directions carefully. Although most test items ask for a single correct answer, some give you the option of marking several choices that are correct.
- First read each question thoroughly. Then look at the choices and try to answer the question.

- Underline key words and phrases in the question. If the question is complicated, try to break it down into small sections that are easy to understand.
- Pay special attention to **qualifiers** such as only, except, etc. For example, negative words in a question can confuse your understanding of what the question asks ("Which of the following is not . . . ").

Qualifier
A word, such as *always, never,* or *often,* that changes the meaning of another word or word group.

If you don't know the answer, eliminate those answers that you know or suspect are wrong. Your goal is to narrow down your choices. Here are some questions to ask:

- Is the choice accurate in its own terms? If there's an error in the choice—for example, a term that is incorrectly defined—the answer is wrong.
- Is the choice relevant? An answer may be accurate, but it may not relate to the essence of the question.
- Are there any qualifiers, such as always, never, all, none, or every? Qualifiers make it easy to find an exception that makes a choice incorrect. For example, the statement that "Children always begin talking before the age of two" can be eliminated as an answer to the question, "When do children generally start to talk?"
- Do the choices give you any clues? Does a puzzling word remind you of a word you know? If you don't know a word, does any part of the word (prefix, suffix, or root) seem familiar to you?

Look for patterns that may lead to the right answer; then use intelligent guessing. Test-taking experts have found patterns in multiple-choice questions that may help you get a better grade. Here is their advice:

- Consider the possibility that a choice that is more general than the others is the right answer.
- Look for a choice that has a middle value in a range (the range can be from small to large, from old to recent). This choice may be the right answer.
- Look for two choices with similar meanings. One of these answers is probably correct.

Make sure you read every word of every answer. Instructors have been known to include answers that are right except for a single word.

When questions are keyed to a long reading passage, read the questions first. This will help you focus on the information you need to answer the questions.

Here are some examples of the kinds of multiple-choice questions you might encounter in an Introduction to Psychology[4] course (the correct answer follows each question):

1. Arnold is at the company party and has had too much to drink. He releases all of his pent-up aggression by yelling at his boss, who promptly fires him. Arnold normally would not have yelled because ________________.
   a. Parties are places where employees are supposed to be able to "loosen up."
   b. Alcohol is a stimulant.

c. Alcohol makes people less concerned with the negative consequences of their behavior.

d. Alcohol inhibits brain centers that control the perception of loudness.

(the correct answer is c)

2. Which of the following has not been shown to be a probable cause of or influence in the development of alcoholism in our society?

a. Intelligence

b. Culture

c. Personality

d. Genetic vulnerability

(the correct answer is a)

True/False Questions

True/false questions test your knowledge of facts and concepts. Read them carefully to evaluate what they truly say. Try to take these questions at face value without searching for hidden meaning. If you're truly stumped, guess (unless you're penalized for wrong answers).

Look for qualifiers in true/false questions—such as all, only, always, because, generally, usually, and sometimes—that can turn a statement that would otherwise be true into one that is false, or vice versa. For example, "The grammar rule, 'I before E except after C,' is always true" is false, whereas "The grammar rule, 'I before E except after C,' is usually true" is true. The qualifier makes the difference. Here are some examples of the kinds of true/false questions you might encounter in an Introduction to Psychology course:

Are the following questions true or false?

1. Alcohol use is always related to increases in hostility, aggression, violence, and abusive behavior. *(False)*
2. Marijuana is always harmful. *(False)*

Essay Questions

An essay question allows you to use writing to demonstrate your knowledge and express your views on a topic. Start by reading the questions and deciding which to tackle (sometimes there's a choice). Then focus on what each question is asking, the mind actions you will have to use, and the writing directions. Read the question carefully, and do everything you are asked to do. Some essay questions may contain more than one part.

Watch for certain action verbs that can help you figure out what to do. Figure 9.2 explains some words commonly used in essay questions. Underline these words as you read any essay question, and use them as a guide.

Next, budget your time and begin to plan. Create an informal outline or think link to map your ideas, and indicate examples you plan to cite to support those ideas. Avoid spending too much time on introductions or flowery prose. Start with a thesis idea or statement that states in a broad way what

Common action verbs on essay tests.

FIGURE 9.2

Analyze—Break into parts and discuss each part separately.
Compare—Explain similarities and differences.
Contrast—Distinguish between items being compared by focusing on differences.
Criticize—Evaluate the positive and negative effects of what is being discussed.
Define—State the essential quality or meaning. Give the common idea.
Describe—Visualize and give information that paints a complete picture.
Discuss—Examine in a complete and detailed way, usually by connecting ideas to examples.
Enumerate/List/Identify—Recall and specify items in the form of a list.
Evaluate—Give your opinion about the value or worth of something, usually by weighing positive and negative effects, and justify your conclusion.
Explain—Make the meaning of something clear, often by making analogies or giving examples.
Illustrate—Supply examples.
Interpret—Explain your personal view of facts and ideas and how they relate to one another.
Outline—Organize and present the subideas or main examples of an idea.
Prove—Use evidence and argument to show that something is true, usually by showing cause and effect or giving examples that fit the idea to be proven
Review—Provide an overview of ideas, and establish their merits and features.
State—Explain clearly, simply, and concisely, being sure that each word gives the image you want.
Summarize—Give the important ideas in brief.
Trace—Present a history of the way something developed, often by showing cause and effect.

your essay will say. As you continue to write your first paragraph, introduce the points of the essay, which may be subideas, causes and effects, or examples. Wrap it up with a concise conclusion.

Use clear, simple language in your essay. Support your ideas with examples, and look back at your outline to make sure you are covering everything. Try to write legibly. If your instructor can't read your ideas, it doesn't matter how good they are. If your handwriting is messy, try printing, skipping every other line, or writing on only one side of the paper.

Do your best to save time to reread and revise your essay after you finish getting your ideas down on paper. Look for ideas you left out and sentences that might confuse the reader. Check for mistakes in grammar, spelling, punctuation, and usage. No matter what subject you are writing about, having a command of these factors will make your work all the more complete and impressive.

Here are some examples of essay questions you might encounter in your Introduction to Psychology course (in each case, notice the action verbs from Figure 9.2):

1. Summarize the theories and research on the uses and effects of daydreaming. Discuss the possible uses for daydreaming in a healthy individual.
2. Describe the physical and psychological effects of alcohol and the problems associated with its use.

Use Specific Techniques for Math Tests

Mathematical test problems present a special challenge to some students, especially those who suffer from math anxiety. These strategies may help you overcome any difficulties you might have:

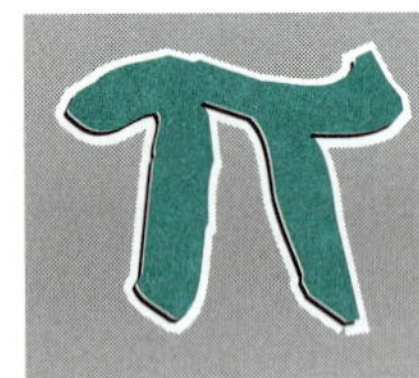

- Analyze problems carefully. Make sure that you take all the "givens" into account as you begin your calculations. Focus also on what you want to find or prove.
- Write down any formulas, theorems, or definitions that apply to the problem. Do this before you begin your calculations.
- Before you begin, estimate to come up with a "ballpark" solution. Then work the problem, and check the solution against your estimate. The two answers should be close. If they're not, recheck your calculations. You may have made a simple calculation error.
- Break the calculation into the smallest possible pieces. Go step-by-step, and don't move on to the next step until you are clear about what you've done so far.
- Recall how you solved similar problems. Past experience can give you valuable clues as to how a particular problem should be handled.
- Draw a picture to help you see the problem. This can be a diagram, a chart, a probability tree, a geometric figure, or any other visual image that relates to the problem at hand.
- Take your time. Precision demands concentration and focus. Also, if you're using a calculator, one wrong keystroke can mean the difference between a right and wrong answer.
- Be neat. When it comes to numbers, mistaken identity can mean the difference between a right and a wrong answer. A 4 that looks like a 9 or a 1 that looks like a 7 can make trouble.
- Use the opposite operation to check your work. When you come up with an answer, work backwards to see if you are right: Use subtraction to check your addition; use division to check multiplication; and so on. Try to check every problem before you hand in your paper.
- Look back at the questions to be sure you did everything that was asked. Did you answer every part of the question? Did you show all the required work? Be as complete as you possibly can.

Knowing a variety of test strategies doesn't guarantee that you will pass the test—you also have to learn the material—but these strategies can help ensure that you are prepared in many ways. If you don't pass a test, do your best to learn from your mistakes.

HOW CAN I LEARN FROM TEST MISTAKES?

The purpose of a test is to see how much you know, not merely to achieve a grade. The knowledge that comes from attending class and studying should allow you to correctly answer test questions. Knowledge also comes when you learn from your mistakes. If you don't learn from what you get wrong on a test, you are likely to repeat the same mistake again on another test and in life. Learn from test mistakes just as you learn from mistakes in your personal life.

> "A little knowledge that acts is worth infinitely more than much knowledge that is idle."
>
> *Kahlil Gibran*

Try to identify patterns in your mistakes by looking for:

- Careless errors. In your rush to complete the exam, did you misread the question or directions, blacken the wrong box, skip a question, or use illegible handwriting?
- Conceptual or factual errors. Did you misunderstand a concept or never learn it in the first place? Did you fail to master certain facts? Did you skip part of the assigned text or miss important classes in which ideas were covered?

You may want to rework the questions you got wrong. Based on the feedback from your instructor, try rewriting an essay, recalculating a math problem, or redoing the questions that follow a reading selection. As frustrating as they are, remember that mistakes show that you are human and that they can help you learn. If you see patterns of careless errors, promise yourself that next time you'll try to budget enough time to double-check your work. If you pick up conceptual and factual errors, rededicate yourself to better preparation.

When you fail a test, don't throw it away. First, take comfort in the fact that a lot of students have been in your shoes and that you are likely to improve your performance. Then recommit to the process by reviewing and analyzing your errors. Be sure you understand why you failed. You may want to ask for an explanation from your instructor. Finally, develop a plan to really learn the material if you didn't understand it in the first place.

You will take many tests during your academic career. Remember that the tests are not a reflection of *you,* but rather a reflection of how well you demonstrated what you learned about a particular subject. As you learn what each instructor expects and as you learn to put these test-taking strategies to use, you will find that test taking may become easier and won't be that dreaded task you used to fear.

EXERCISES AND ACTIVITIES

9.1 Test Analysis

When you get back your next test, take a detailed look at your performance:

- Start by writing what you think of your test performance and grade. Were you pleased or disappointed? If you made errors, were they careless or due to not knowing facts and concepts?
- Next, list the test preparation activities that helped you do well on the exam and the activities you wish you had done—and intend to do for the next exam:

  Positive actions I took:

  Positive actions I intend to take next time:
- Finally, list the activities you don't intend to repeat when studying for the next test:

9.2 Learning from My Mistakes

For this exercise, use an exam on which you made one or more mistakes:

- Why do you think you answered the question(s) incorrectly?
- Did any qualifying terms—such as always, sometimes, never, often, occasionally, only, no, and not—make the question(s) more difficult or confusing? What steps could you have taken to clarify the meaning?
- Did you try to guess the correct answer? If so, why do you think you made the wrong choice?
- Did you feel rushed? If you had had more time, do you think you would have gotten the right answer(s)? What could you have done to budget your time more effectively?
- If an essay question was a problem, what do you think went wrong? What will you do differently the next time you face an essay question on a test?

9.3 Create a Test

Choose a chapter of this text, and practice predicting possible test questions. Write at least 3 true/false questions, 3 multiple-choice questions, and 3 essay questions.

Endnotes

1 Sheila Tobias, *Overcoming Math Anxiety* (New York: W. W. Norton & Company, 1993), 50.

2 Ibid.

3 Ibid.

4 Many of the examples of objective questions found in this chapter are from Gary W. Piggrem, "Test Item File for Charles G. Morris." In *Understanding Psychology,* 3rd ed. (Upper Saddle River, NJ: Prentice Hall, 1996).

PLANNING

PLANNING

10 Managing Time

How often have you used that old excuse, "I didn't have time (to study for my test or to read the chapters or to prepare my speech)"? In our all too busy lives, it is easy to make that statement. However, the truth is, you probably did have time; you simply chose not to use your time to focus on that activity.

Time is one of your most valuable and precious resources. Unlike money or opportunity or connections, time doesn't discriminate—everyone has the same twenty-four hours in a day, every day. Your responsibility and your potential for success lie in how you use yours. You cannot manipulate or change how time passes, but you can spend it taking steps to achieve your goals.

In this chapter, you will explore answers to the following questions:

- How can I manage my time?
- What time management strategies can I try?
- Why is procrastination a problem?

HOW CAN I MANAGE MY TIME?

Time management, like physical fitness, is a lifelong pursuit. No one can plan a perfect schedule or build a terrific physique and then be done. You'll work at time management throughout your life, and it can be tiring. Your ability to manage your time will vary with your mood, your stress level, your activity level, and other factors. You're human; don't expect perfection. Just do your best. Time management involves taking responsibility for how you spend your time, building a schedule, and making your schedule work through lists and other strategies.

Taking Responsibility for How I Spend My Time

Being in control of how you manage your time is a key factor in taking responsibility for yourself and your choices. When you plan your activities with an eye toward achieving your most important goals, you are taking personal responsibility for how you live. Life changes and the judgments of others are among the factors that can affect your control.

Life Changes

Life's sudden changes and circumstances often make you feel out of control. One minute you seem to be on track, and the next minute chaos hits: Your car breaks down; your relationship falls apart; you fail a class; you develop a medical problem. Coping with all of these changes can cause stress. As your stress level rises, your sense of control dwindles.

Although you cannot always choose your circumstances, you might be able to choose how to handle them. Dr. Stephen Covey says that language is important in trying to take action. Using language like "I have to" and "They made me" robs you of personal power. For example, saying that "I have to go to school" or "I have to work part-time" can make you feel that others control your life. However, language like "I have decided to" and "I prefer" helps energize your power to choose. Then you can turn "I have to go to school" into "I prefer to go to school rather than end up working full-time in a dead-end job."

Judgments of Others

The judgments of others can also intimidate you into not taking responsibility for your time. A student who feels no one will hire him because of his weight may not search for jobs. A student who feels her instructor is prejudiced against her might not study for that instructor's course. Try not to let these barriers rob you of your control of your time.

Early in his life, Malcolm X was told that he had no business aspiring to be a lawyer in spite of his excellent record as a student. He was constantly demeaned because of his race. However, he did not let the ignorance of others stand in his way.

Instead of giving in to judgments, try to choose actions that improve your circumstances. If you lose a job, spending an hour a day investigating other job opportunities is a better use of your time than watching TV. If you have trouble with an instructor, you can address the problem with that instructor directly and try to make the most of your time in the course. If that doesn't work, you could change classes, spend that time in other important pursuits, or retake the course in summer school while working part-time. Try to find an option that will allow you to be in control of your time.

Time can be your ally if you make smart choices about how to use it. Building a schedule can help you decide when to accomplish the activities you choose.

Building a Schedule

Just as a road map helps you travel from place to place, a schedule is a time-and-activity map that helps you get from the beginning of the day (or week or month) to the end as smoothly as possible. A written schedule helps you gain control of your life. Schedules have two major advantages: They allocate segments of time for the fulfillment of your daily, weekly, monthly, and longer-term goals; and they serve as a concrete reminder of tasks, events, due dates, responsibilities, and deadlines. Few moments are more stressful than suddenly realizing you have forgotten to write a paper, take a test, or be at work. Scheduling can help you avoid events like these.

"Efficiency is intelligent laziness."
David Dunham

Keep a Date Book

Gather the tools of the trade: a pen or pencil and a date book (sometimes called a planner). Some of you already have date books and may have used them for years. Others may have had no luck with them or have never tried. Even if you don't feel you are the type of person who would use one, give it a try. A date book is indispensable for keeping track of your time. Paul Timm says, "Most time management experts agree that rule number one in a thoughtful planning process is: Use some form of a planner where you can write things down."[1]

There are two major types of date books. The day-at-a-glance version devotes a page to each day. It gives you ample space to write the day's activities but makes it difficult to see what's ahead. The week-at-a-glance book gives you a view of the week's plans but has less room to write per day. If you write out your daily plans in detail, you might like the day-at-a-glance version. If you prefer to remind yourself of plans ahead of time, try the book that shows a week's schedule all at once. Some date books contain additional sections that allow you to note plans and goals for the year as a whole and for each month. You can also create your own sheets for yearly and monthly notations in a notepad section, if your book has one, or on plain paper that you can then insert into the book.

Another option to consider is an electronic planner. These are compact minicomputers that can hold a large amount of information. You can use them to schedule your days and weeks, make to-do lists, and create and store an address book. Electronic planners are powerful, convenient, and often fun. On the other hand, they certainly cost more than the paper version, and you can lose a lot of important data if something goes wrong with the computer. Evaluate your options, and decide what you like best.

Set Weekly and Daily Goals

The most ideal time management starts with the smallest tasks and builds to bigger ones. Setting short-term goals that tie in to your long-term goals lends the following benefits:

- Increased meaning for your daily activities
- Shaping your path toward the achievement of your long-term goals
- A sense of order and progress

For students as well as working people, the week is often the easiest unit of time to consider at one shot. Weekly goal setting and planning allow you to keep track of day-to-day activities while giving you the larger perspective of what is coming up during the week. Take some time before each week starts to remind yourself of your long-term goals. Keeping long-term goals in mind will help you determine related short-term goals you can accomplish during the week to come.

Figure 10.1 shows parts of a daily schedule and a weekly schedule.

Making My Schedule Work

Link Daily and Weekly Goals with Long-Term Goals

After you evaluate what you need to accomplish in the coming year, semester, month, week, and day in order to reach your long-term goals, use your schedule to record those steps. Write down the short-term goals that will enable you to stay on track. Here is how a student might map out two different goals over a year's time:

> "The right time is any time that one is still so lucky as to have. . . .Live!"
> *Henry James*

This year:	Complete enough courses to graduate. Earn a spot on the track team.
This semester:	Complete my accounting class with a B average or higher. Begin a fitness plan that includes exercising daily.
This month:	Set up study group schedule to coincide with quizzes. Begin running and weight lifting.
This week:	Meet with study group; go over material for Friday's quiz. Go for a run three times; go to weight room twice.
Today:	Go over Chapter 3 in accounting text. Run for 40 minutes.

Prioritize Goals

Prioritizing enables you to use your date book with maximum efficiency. On any given day, your goals will have varying degrees of importance. Record your goals first, and then label them according to level of importance, using these categories: priority 1, priority 2, and priority 3. Identify these categories using any code that makes sense to you. Some people use numbers, as above; some use letters (A, B, C). Some write activities in different colors according to priority level, and others use symbols (*, +, –).

- *Priority 1* activities are the most important things in your life. They may include attending class, going to a part-time job, putting gas in the car, and paying bills.

Daily and weekly schedules.

MONDAY, MARCH 27		2000
TIME	TASKS	PRIORITY
7:00 AM		☆
8:00	Up at 8am – finish homework	
9:00		
10:00	Business Administration	☆
11:00	Quiz	
12:00 PM		
1:00	Lunch	☆
2:00	Writing Seminar (peer editing today)	
3:00		
4:00	check on Ms. Schwartz's office hrs.	
5:00	5:30 work out	
6:00	↳6:30	
7:00	Dinner	
8:00	Read two chapters for Business Admin.	
9:00	↓	
10:00		
11:00		

Monday, March 27

8			CALL: Maggie Blair
9	Biology	1	Financial Aid Office
10		2	THEATRE ☆Paramedic role-play ☆
11	Algebra II	3	
12		4	
Evening	6pm yoga class	5	

Tuesday, March 28

8	Finish reading assignment!		Work @ library
9		1	
10	ENG III	2	(study for quiz)
11	↓	3	
12		4	
Evening		5	↓ until 7pm

Wednesday, March 29

8			Meet w/advisor
9	Biology	1	
10		2	Theatre
11	Algebra II	3	
12		4	☆ Pick up photos
Evening	6pm Aerobics	5	

- *Priority 2* activities are part of your routine. Examples include running errands, working out, participating in a school organization, or cleaning your room. Priority 2 tasks are important but more flexible than priority 1s.
- *Priority 3* activities are those you would like to do but can reschedule without much sacrifice. Examples might be a trip to the mall, a visit to a friend, a social phone call, a sports event, a movie, or a hair appointment. As much as you would like to accomplish them, you don't consider them urgent. Many people don't enter priority 3 tasks in their date books until they are sure they have time to get them done.

Prioritizing your activities is essential for two reasons. First, some activities are more important than others, and effective time management requires that you focus most of your energy on priority 1 items. Second, looking at all your priorities helps you plan when you can get things done. Often, it's not possible to get all your priority 1 activities done early in the day, especially if these activities involve scheduled classes or meetings. Prioritizing helps you set priority 1 items and then schedule priority 2 and 3 items around them as they fit.

Keep Track of Events

Your date book also enables you to schedule events. Rather than thinking of events as separate from goals, tie them to your long-term goals just as you would your other tasks. For example, attending a wedding in a few months contributes to your commitment to spending time with your family. Being aware of quiz dates, due dates for assignments, and meeting dates will aid your goals to achieve in school and become involved. Note events in your date book so that you can stay aware of them ahead of time. Write them in daily, weekly, monthly, or even yearly sections where a quick look will remind you that they are approaching. Writing them down will also help you see where they fit in the context of all your other activities. For example, if you have three big tests and a presentation all in one week, you'll want to take time in the weeks beforehand to prepare for them all.

Following are some kinds of events worth noting in your date book:

- Due dates for papers, projects, presentations, and tests
- Important meetings, medical appointments, or due dates for car payments
- Birthdays, anniversaries, social events, holidays, and other special occasions
- Benchmarks for steps toward a goal, such as due dates for sections of a project or a deadline for losing five pounds on your way to twenty

List Low-Priority Goals Separately

Priority 3 tasks can be hard to accomplish. As the least important tasks, they often get pushed off from one day to the next. You may spend valuable time rewriting these items day after day in your date book instead of getting them done. One solution is to keep a list of priority 3 tasks in a separate place in your date book. That way, when you have an unexpected pocket of free time, you can consult your list and see what you have time to accomplish—making a trip to the post office, writing a card, returning a borrowed tape, or giving some clothes to charity. Keep this list current by crossing off items as you accomplish them and writing in new items as soon as you think of them. Rewrite the list when it gets too messy.

WHAT TIME MANAGEMENT STRATEGIES CAN I TRY?

Managing time takes thought and energy. Here are some additional strategies to try:

1. Plan your schedule each week. Before each week starts, note events, goals, and priorities. Look at the map of your week to decide where to fit activities like studying and priority 3 items. For example, if you have a test on Thursday, you can plan study sessions on the days up until then. If you have more free time on Tuesday and Friday than on other days, you can plan workouts or priority 3 activities at those times. Looking at the whole week will help you avoid being surprised by something you had forgotten was coming up.

2. Make and use to-do lists. Use a to-do list to record the things you want to accomplish. If you generate a daily or weekly to-do list on a separate piece of paper, you can look at all tasks and goals at once. This will help you consider time frames and priorities. You might want to prioritize your tasks and transfer them to appropriate places in your date book. Some people create daily to-do lists right on their date book pages. You can tailor a to-do list to an important event such as exam week or an especially busy day when you have a family gathering or a presentation to make. This kind of specific to-do list can help you prioritize and accomplish an unusually large task load.

3. Make thinking about time a priority. Timm recommends that you devote a minimum of 10 to 15 minutes a day to planning your schedule. Although making a schedule takes time, it can mean hours of saved time later. Say you have two errands to run, both on the other side of town; not planning ahead could result in your driving across town twice in one day. The extra driving time is far more than it would have taken to plan the day in advance.

4. Refer to your schedule. Many people make detailed schedules, only to forget to look at them. Carry your date book wherever you go, and check it throughout the day. Find a date book size you like—there are books that fit into your backpack, bag, or even your pocket.

5. Post monthly and yearly calendars at home. Keeping a calendar on the wall will help you stay aware of important events. You can purchase one or draw it yourself, month by month, on plain paper. Use a yearly or a monthly version (Figure 10.2 shows part of a monthly calendar), and keep it where you can refer to it often. You can also make the calendar a family project so that you stay aware of each other's plans. Knowing each other's schedules can also help you avoid scheduling problems such as two people needing the car at the same time.

6. Schedule downtime. When you're wiped out from too much activity, you don't have the energy to accomplish much with your time. A little downtime will refresh you and improve your attitude. Even half an hour a day will help. Fill the time with whatever relaxes you—having a snack, reading, watching TV, playing a game or sport, walking, writing, or just doing nothing. Make downtime a priority.

7. Be flexible. Because priorities determine the map of your day, week, month, or year, any priority shift can jumble your schedule. Be ready to reschedule your tasks as your priorities change. On Monday, a homework assignment due in a week might be priority 2. By Saturday, it has become priority 1. On some days, a surprise priority (such as a medical emergency or a family situation) may pop up and force you to cancel everything else on your schedule. Other days, a sports practice may be canceled, and you will have extra time on your hands. Adjust to whatever each day brings.

FIGURE 10.2 Monthly calendar.

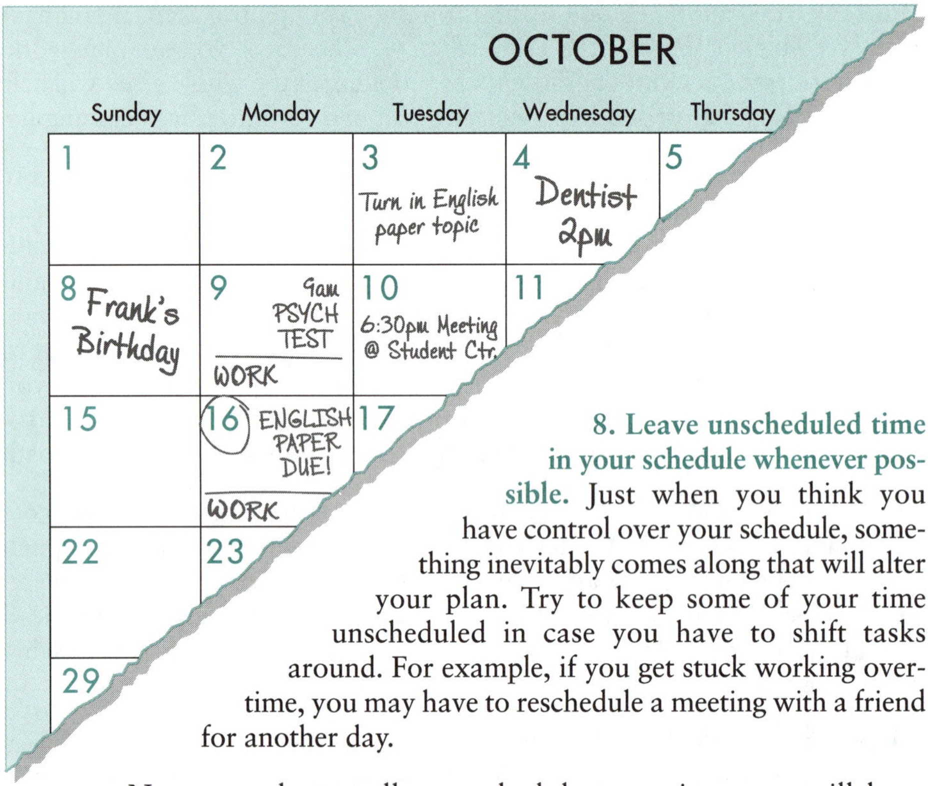

Sunday	Monday	Tuesday	Wednesday	Thursday
1	2	3 Turn in English paper topic	4 Dentist 2pm	5
8 Frank's Birthday	9 9am PSYCH TEST / WORK	10 6:30pm Meeting @ Student Ctr.	11	
15	16 ENGLISH PAPER DUE! / WORK	17		
22	23			
29				

8. **Leave unscheduled time in your schedule whenever possible.** Just when you think you have control over your schedule, something inevitably comes along that will alter your plan. Try to keep some of your time unscheduled in case you have to shift tasks around. For example, if you get stuck working overtime, you may have to reschedule a meeting with a friend for another day.

No matter how well you schedule your time, you will have moments when it's hard to stay in control. Knowing how to identify and avoid procrastination and other time traps will help you get back on track.

WHY IS PROCRASTINATION A PROBLEM?

Procrastination
The act of putting off something that needs to be done.

Procrastination occurs when you postpone unpleasant or burdensome tasks. People procrastinate for different reasons. Having trouble with goal setting is one reason. People may project goals too far into the future, set unrealistic goals that are too frustrating to reach, or have no goals at all. People also procrastinate because they don't believe in their ability to complete a task or don't believe in themselves in general. As natural as these tendencies are, they can also be extremely harmful. If continued over a period of time, procrastination can develop into a habit that will dominate a person's behavior. Following are some ways to face your tendencies to procrastinate and just do it!

Strategies to Fight Procrastination

The following hints will help you fight your tendencies to procrastinate.

- **Weigh the benefits (to you and others) of completing the task versus the effects of procrastinating.** What rewards lie ahead if you get it done? A burden off your shoulders? Some free time? Better grades? What will be the effects if you continue to put it off? Which situation has better effects? Chances are, you will benefit more in the long term from facing the task head-on.

- Set reasonable goals. Plan your goals carefully, allowing enough time to complete them. Unreasonable goals can be so intimidating that you do nothing at all. "Pay off the credit-card bill next month" could throw you. However, "Pay off the credit-card bill in six months" might inspire you to take action.

- Get started. Going from doing nothing to doing something is often the hardest part of avoiding procrastination. You might want to use the motivation techniques from Chapter 1 to help you take the first step. Once you start, you may find it easier to continue.

- Break the task into smaller parts. If it seems overwhelming, look at the task in terms of its parts. How can you approach it step-by-step? If you can concentrate on achieving one small goal at a time, the task may become less of a burden.

- Ask for help with tasks and projects at school, work, and home. You don't always have to go it alone. Instructors, supervisors, and family members can lend support, helping you to complete a dreaded task. For example, if you have put off an intimidating assignment, ask your instructor for guidance. If you avoid a project because you dislike the person with whom you have to work, talk to your instructor about adjusting the assignment of tasks. If you need accommodations due to a disability, don't assume that others know about it. Once you identify what's holding you up, see who can help you face the task.

- Don't expect perfection. No one is perfect. Being able to do something flawlessly is not a requirement for trying. Most people learn by starting at the beginning and wading through plenty of mistakes and confusion. It's better to try your best than to do nothing at all.

- Consider how you would operate if you were looking forward to something you really wanted to do. You might not be late if you were headed to the airport for a flight to the Bahamas! See if you can transfer that behavior to a task that isn't quite as much fun.

Procrastination is natural, but it can cause you problems if you let it get the best of you. When it does happen, take some time to think about the causes. What is it about this situation that frightens you or puts you off? Answering that question can help you address what causes lie underneath the procrastination. These causes might indicate a deeper problem that needs to be solved.

Other Time Traps to Avoid

Procrastination isn't the only way to spend your time in less than productive ways. Keep an eye out for these situations, too:

- Saying yes when you really don't have the time. Many people, in their efforts to please others, agree to help with tasks they can't easily fit into their schedule. Being reliable is great, but not when it is at your own expense. Learn to say no when you need to. First, resist the desire to respond right away. Then ask yourself what effects a new responsibility will have on your schedule. Be honest with yourself about whether you have the time to make a new commitment. If it will cause you more trouble than it seems to be worth, say no graciously.

- Studying at a bad time of day. At what point in the day do you have the most energy? Is that when you study? If not, you may be wasting time. When you are tired, you may need extra time to fully understand your material. If you study when you are most alert, you will be able to take in more information in less time.

- Studying in a distracting location. Find an environment that helps you maximize study time. If you need to be alone to concentrate, for example, studying near family members might interfere with your focus. Conversely, people who require a busier environment to stay alert might need to choose a more active setting.

- Not thinking ahead. Forgetting important things is a big time drain. One book left at home can cost you extra time going back and forth. One forgotten phone call can mean you have to do what you wanted to ask someone else to do. Five minutes of scheduling in the morning or the night before can save you hours.

- Not curbing your social time. Time passes quickly when you're having fun. You plan to make a quick telephone call and the next thing you know you've been talking for an hour, losing time you could have used for studying or sleep. Don't cut out all socializing, but wear a watch and stay aware of the time. If friends invite you to a ballgame one evening and you know you can't spend a whole evening out, consider joining them after the game. Your friends will most likely respect your priorities, and you will respect yourself when you see the rewards.

- Pushing yourself too far. You've probably experienced one of those study sessions during which, at a certain point, you realize that you haven't absorbed anything for the last hour. Sometimes you just need a break. Stay aware of your energy level; when you just can't seem to concentrate anymore, take a refresher—stretch, get a drink or a snack, go for a walk, take a nap. You're much better off using some of your time to revive yourself rather than trying in vain to focus.

As you learn to manage your time effectively and efficiently, think about how time is like money. You can spend it foolishly and have nothing to show for it, or you can invest your time wisely and reap big rewards.

EXERCISES AND ACTIVITIES

Short-Term Scheduling 10.1

Take a close look at your schedule for the coming month, including events, important dates, and steps toward goals. On a calendar or day planner, fill in the name of the month and appropriate numbers for the days. Then record what you hope to accomplish, including the following:

- Due dates for papers, projects, and presentations
- Test dates
- Important meetings, medical appointments, and due dates for car payments
- Birthdays, anniversaries, and other special occasions
- Steps toward long-term goals

This kind of chart will help you see the big picture of your month. To stay on target from day to day, check these dates against the entries in your date book, and make sure that they are indicated there as well.

Discover How I Spend My Time 10.2

In the chart to the right, estimate the total time you think you spend per week on each listed activity. Then add the hours together. If your number is over 168 (the number of hours in a week), rethink your original estimates, and recalculate the total so that it equals or is below 168.

When your estimate is at or below 168, subtract that number (the total number of hours you estimate you spend on these activities) from 168. Whatever is left over is your estimate of hours that you spend in unscheduled activities.

168

Minus total __________

Unscheduled time __________

ACTIVITY	ESTIMATED TIME SPENT
Class	
Work	
Studying	
Sleeping	
Eating	
Family time	
Commuting/traveling	
Chores and personal business	
Friends and important relationships	
Telephone time	
Leisure/entertainment	
Spiritual life	
Total	

Now spend a week recording exactly how you spend your time. The following chart has blocks showing half-hour increments. As you go through the week, write in what you do each hour, indicating when you started and when you

MONDAY		TUESDAY		WEDNESDAY		THURSDAY	
Time	Activity	Time	Activity	Time	Activity	Time	Activity
5:00 AM		5:00 AM		5:00 AM		5:00 AM	
5:30 AM		5:30 AM		5:30 AM		5:30 AM	
6:00 AM		6:00 AM		6:00 AM		6:00 AM	
6:30 AM		6:30 AM		6:30 AM		6:30 AM	
7:00 AM		7:00 AM		7:00 AM		7:00 AM	
7:30 AM		7:30 AM		7:30 AM		7:30 AM	
8:00 AM		8:00 AM		8:00 AM		8:00 AM	
8:30 AM		8:30 AM		8:30 AM		8:30 AM	
9:00 AM		9:00 AM		9:00 AM		9:00 AM	
9:30 AM		9:30 AM		9:30 AM		9:30 AM	
10:00 AM		10:00 AM		10:00 AM		10:00 AM	
10:30 AM		10:30 AM		10:30 AM		10:30 AM	
11:00 AM		11:00 AM		11:00 AM		11:00 AM	
11:30 AM		11:30 AM		11:30 AM		11:30 AM	
12:00 PM		12:00 PM		12:00 PM		12:00 PM	
12:30 PM		12:30 PM		12:30 PM		12:30 PM	
1:00 PM		1:00 PM		1:00 PM		1:00 PM	
1:30 PM		1:30 PM		1:30 PM		1:30 PM	
2:00 PM		2:00 PM		2:00 PM		2:00 PM	
2:30 PM		2:30 PM		2:30 PM		2:30 PM	
3:00 PM		3:00 PM		3:00 PM		3:00 PM	
3:30 PM		3:30 PM		3:30 PM		3:30 PM	
4:00 PM		4:00 PM		4:00 PM		4:00 PM	
4:30 PM		4:30 PM		4:30 PM		4:30 PM	
5:00 PM		5:00 PM		5:00 PM		5:00 PM	
5:30 PM		5:30 PM		5:30 PM		5:30 PM	
6:00 PM		6:00 PM		6:00 PM		6:00 PM	
6:30 PM		6:30 PM		6:30 PM		6:30 PM	
7:00 PM		7:00 PM		7:00 PM		7:00 PM	
7:30 PM		7:30 PM		7:30 PM		7:30 PM	
8:00 PM		8:00 PM		8:00 PM		8:00 PM	
8:30 PM		8:30 PM		8:30 PM		8:30 PM	
9:00 PM		9:00 PM		9:00 PM		9:00 PM	
9:30 PM		9:30 PM		9:30 PM		9:30 PM	
10:00 PM		10:00 PM		10:00 PM		10:00 PM	
10:30 PM		10:30 PM		10:30 PM		10:30 PM	
11:00 PM		11:00 PM		11:00 PM		11:00 PM	
11:30 PM		11:30 PM		11:30 PM		11:30 PM	

FRIDAY Time	Activity	SATURDAY Time	Activity	SUNDAY Time	Activity	NOTES for the week of ———
5:00 AM		5:00 AM		5:00 AM		
5:30 AM		5:30 AM		5:30 AM		
6:00 AM		6:00 AM		6:00 AM		
6:30 AM		6:30 AM		6:30 AM		
7:00 AM		7:00 AM		7:00 AM		
7:30 AM		7:30 AM		7:30 AM		
8:00 AM		8:00 AM		8:00 AM		
8:30 AM		8:30 AM		8:30 AM		
9:00 AM		9:00 AM		9:00 AM		
9:30 AM		9:30 AM		9:30 AM		
10:00 AM		10:00 AM		10:00 AM		
10:30 AM		10:30 AM		10:30 AM		
11:00 AM		11:00 AM		11:00 AM		
11:30 AM		11:30 AM		11:30 AM		
12:00 PM		12:00 PM		12:00 PM		
12:30 PM		12:30 PM		12:30 PM		
1:00 PM		1:00 PM		1:00 PM		
1:30 PM		1:30 PM		1:30 PM		
2:00 PM		2:00 PM		2:00 PM		
2:30 PM		2:30 PM		2:30 PM		
3:00 PM		3:00 PM		3:00 PM		
3:30 PM		3:30 PM		3:30 PM		
4:00 PM		4:00 PM		4:00 PM		
4:30 PM		4:30 PM		4:30 PM		
5:00 PM		5:00 PM		5:00 PM		
5:30 PM		5:30 PM		5:30 PM		
6:00 PM		6:00 PM		6:00 PM		
6:30 PM		6:30 PM		6:30 PM		
7:00 PM		7:00 PM		7:00 PM		
7:30 PM		7:30 PM		7:30 PM		
8:00 PM		8:00 PM		8:00 PM		
8:30 PM		8:30 PM		8:30 PM		
9:00 PM		9:00 PM		9:00 PM		
9:30 PM		9:30 PM		9:30 PM		
10:00 PM		10:00 PM		10:00 PM		
10:30 PM		10:30 PM		10:30 PM		
11:00 PM		11:00 PM		11:00 PM		
11:30 PM		11:30 PM		11:30 PM		

stopped. Don't forget activities that don't feel like activities, such as sleeping, relaxing, and watching TV. Also, beware of recording how you want to spend your time or how you think you should have spent your time; be perfectly honest about your schedule. There are no wrong answers.

Now go through this chart and look at how many hours you actually spent on the activities for which you estimated your hours before. Tally the hours in the boxes in the following table using straight tally marks; round off to half-hours, and use a short tally mark for a half-hour spent. At the far right of the table, total the hours for each activity.

ACTIVITY	TIME TALLIED OVER ONE-WEEK PERIOD	TOTAL TIME IN HOURS
Example: Class	𝍸 𝍸 𝍸 ||	16.5
Class		
Work		
Studying		
Sleeping		
Eating		
Family time		
Commuting/traveling		
Chores and personal business		
Friends and important relationships		
Telephone time		
Leisure/entertainment		
Spiritual life		

Add the totals on the right to find your GRAND TOTAL: ____________

Compare your grand total to your estimated grand total; compare your actual activity hour totals to your estimated activity hour totals. What matches and what doesn't? Describe the similarities and differences:

What is the one biggest surprise about how you spend your time?

Name one change you would like to make in how you spend your time:

Think about what kinds of changes you could make that will help you improve your ability to set and achieve goals. Ask yourself important questions about what you do daily, weekly, and monthly. On what activities do you think you should spend more or less time? For this last chart, write the hours for each activity that represent your ideal week.

ACTIVITY	IDEAL TIME IN HOURS
Class	
Work	
Studying	
Sleeping	
Eating	
Family time	
Commuting/traveling	
Chores and personal business	
Friends and important relationships	
Telephone time	
Leisure/entertainment	
Spiritual life	
Total	

To-Do Lists 10.3

Make a to-do list for what you have to do tomorrow. Include all tasks—priority 1, 2, and 3—and events.

- What are the effects of your procrastination? Discuss how procrastination may affect the quality of your work, motivation, productivity, ability to be on time, grades, or self-perception:

- Think about a specific time when you procrastinated. Describe what happened:

- Choose one of your procrastination habits to work on changing. What do you plan to work on, and how?

- When you get a chance to work on it, describe what happened. Did you make progress? How did it help you to fight your procrastination?

Endnotes

1 Paul R. Timm, *Successful Self-Management: A Psychologically Sound Approach to Personal Effectiveness* (Los Altos, CA: Crisp Publications, 1987), 22–41.

EXPRESS

EXPRESS

11

Communicating

Words, joined to form ideas, are tools that have enormous power. Whether you write an essay, a memo to an admissions counselor, or a love letter over e-mail, words allow you to take your ideas out of the realm of thought and give them a form that other people can read or listen to and consider. You can harness their power for your own purposes.

Two types of successful communication will be a factor in your success in life: written communication and oral communication. Set a goal for yourself: Strive continually to improve your knowledge of how to use words to construct understandable ideas and accomplish tasks efficiently.

In this chapter, you will explore answers to the following questions:

- How can I express myself effectively?
- Why do good speaking skills matter?
- Why does good writing matter?
- What are the elements of effective writing?

HOW CAN I EXPRESS MYSELF EFFECTIVELY?

The only way for people to know each other's needs is to communicate as clearly and directly as possible. Successful communication promotes successful school, work, and personal relationships. Exploring communication styles, addressing specific communication problems, and using specific success strategies will help you express yourself effectively.

The Styles

Even though the communication process is the same for all individuals, different people have different styles of communicating. Problems arise when one person has trouble "translating" a message that comes from someone who uses a different style. There are at least four communication styles into which people tend to fit: the intuitor, the senser, the thinker, and the feeler. Of course, people may shift around or possess characteristics from more than one category, but for most people, one or two styles are dominant. Recognizing specific styles in others will help you communicate more clearly.[1]

The four styles are derived from the Myers-Briggs Type Indicator® (MBTI). Not all individual learning styles within the assessments are mentioned, and the styles that are noted may correspond with different styles in different situations, but these matchups depict the most common associations. Finding where your learning styles fit in this list may help you to determine your dominant communication style or styles:

FOUR COMMUNICATION STYLES

1. A person using the *intuitor* style is interested in ideas more than details, often moves from one concept or generalization to another without referring to examples, values insights and revelations, talks about having a vision, looks toward the future, and can be oriented toward the spiritual.
2. A person showing the style of *senser* prefers details or concrete examples to ideas and generalizations, is often interested in the parts rather than the whole, prefers the here and now to the past or future, is suspicious of sudden insights or revelations, and feels that seeing is believing.
3. A person using the *thinker* style prefers to analyze situations, likes to solve problems logically, sees ideas and examples as useful if they help to figure something out, and becomes impatient with emotions or personal stories unless they have a practical purpose.
4. A person showing the style of *feeler* is concerned with ideas and examples that relate to people, often reacts emotionally, is concerned with values and their effects on people and other living things, and doesn't like cold logic or too much detail.

You can benefit from shifting from style to style according to the situation, particularly when trying to communicate with someone who prefers a style different from yours. Shifting, however, is not always easy or possible. The most important task is to try to understand the different styles and to help others understand yours. In general, no one style is any better than another. Each has its own positive effects that enhance communication and negative effects that can hinder it, depending on the situation.

Adjusting to the Listener's Style

When you are the speaker, you will benefit from an understanding of both your own style and the styles of your listeners. It doesn't matter how clear you think you are being if the person you are speaking to can't "translate" your message by understanding your style. Try to take your listener's style into consideration when you communicate.

Following is an example of how adjusting to the listener can aid communication:

> An intuitor-dominant instructor to a senser-dominant student: "Your writing isn't clear."
>
> The student's reply: "What do you mean?"

WITHOUT ADJUSTMENT

- If the intuitor doesn't take note of the senser's need for detail and examples, he or she may continue with a string of big-picture ideas that might further confuse and turn off the senser. "You need to elaborate more. Try writing with your vision in mind. You're not considering your audience."

WITH ADJUSTMENT

- If the intuitor shifts toward a focus on detail and away from his or her natural focus on ideas, the senser may begin to understand, and the lines of communication can open. "You introduced your central idea at the beginning but then didn't really support it until the fourth paragraph. You need to connect each paragraph's idea to the central idea. Also, not using a lot of examples for support makes it seem as though you are writing to a very experienced audience."

Adjusting to the Communicator's Style

As a facet of communication, listening is just as important as speaking. When you are the listener, try to stay aware of the communication style of the person who is speaking to you. Observe how that style satisfies, or doesn't satisfy, what a person of your particular style prefers to hear. Work to understand the speaker in the context of his or her style, and translate the message into one that makes sense to you. Following is an example of how adjusting to the communicator can boost understanding:

> A feeler-dominant employee to a thinker-dominant supervisor: "I'm really upset about how you've talked down to me. I don't think you've been fair. I haven't been able to concentrate since our discussion, and it's hurting my performance."

WITHOUT ADJUSTMENT

- If the thinker becomes annoyed with the feeler's focus on emotions, he or she may ignore them, putting up an even stronger barrier between the two people. "There's no reason to be upset. I told you clearly and specifically what needs to be done. There's nothing else to discuss."

WITH ADJUSTMENT

- If the thinker considers that emotions are dominant in the feeler's perspective, he or she could respond to those emotions in a way that still searches for the explanations and logic the thinker understands best: "Let's talk about how you feel. Please explain to me what has caused you to become upset, and we'll discuss how we can improve the situation."

As you learn to adjust your speaking style, you'll discover the benefits of why good speaking matters.

WHY DO GOOD SPEAKING SKILLS MATTER?

"You can complain because roses have thorns, or you can rejoice that thorns have roses."
Ziggy

If you are able to express yourself in conversations and interviews, you will always have an advantage over someone who cannot speak clearly and effectively. When a person can speak with confidence in a variety of settings and situations, that person appears to be confident and self-assured to others. This results in the speaker getting more respect from the listener. Communicating well is a skill you will always use, and it is especially important now that you are transitioning from high school to college. As you work through all the decisions you are facing, you will need to communicate clearly with many people. And, as you enter the college environment, you will find yourself needing to clearly communicate with instructors and roommates.

Communicating with Others Now

As you prepare for college, you are probably discovering that you have to practice the art of communication now with many individuals—college representatives, high school counselors, and even parents.

College Representatives

When college recruiters come to your school, they are, of course, selling you on their school. They want you to attend the college they represent! However, college recruiters may also be seeing if you can sell yourself to them. Students should be able to ask pertinent questions about the college and should be able to tell the recruiter some information about themselves. These communication skills can help focus the discussion, so you can get the most out of your interview with the recruiter.

You may also find yourself having to make numerous phone calls to individuals at the college. For example, you may need to call and discuss housing arrangements with someone and scholarships with someone else. Without good communication skills, you might end up in a residential hall you don't like or without a scholarship at all! The following tips will help you communicate clearly when talking over the telephone:

- Identify yourself. State your name, and tell the person you a are prospective student.
- Identify your reason for calling. For example, tell the person you would like more information on the procedure for applying for scholarships.

11

- Write down the information. Using a shorthand you will understand later, take down what you find out.
- Ask questions. You should ask questions that you have prepared and written down prior to calling.
- Ask for clarification. When you don't understand something, make sure to clarify it before you hang up.
- Repeat important information. The person has given you information; repeat the main points before hanging up. This gives you a chance to look over your notes and to make sure you've understood and written down the information correctly.
- Follow up. Rewrite your notes if necessary, and complete any other tasks associated with the call.

High School Counselors

During the last year of high school, you should be discussing college options with your high school counselors. Everyone's experience with high school counselors is different, but with effective communication skills, you can be in control and get the most out of your meetings. If you are in a large school, your meetings with the counselor may be limited; without good communication, you might not get the information you need. However, if you represent yourself well through your communication skills, your counselor may make more of an effort to listen to your concerns and respond accordingly, see you multiple times, help you with your decisions, and pursue answers to your questions. The following tips can help you communicate clearly with your high school counselor:

- Keep your scheduled appointments and be on time.
- Identify your reason for making the appointment. Again, this will help focus your time. For example, if you don't know how to get started, your counselor can help you. If you've already done some background work and have narrowed your choices, your counselor can help you at that point in the process.
- Come prepared. If there are tasks you can complete prior to your meeting, do so! You can use your time better if you have done your homework.
- Express your concerns and desires. Be clear and honest about what you want to get out of the meeting or about what you want to do with your future.
- Ask questions. If you have never thought about college before and don't even know what to ask, tell the counselor! He or she will give you information to get you started.
- Write down information. Be very careful to write down deadlines and important dates. If the counselor gives you a handout or flier, be sure to highlight the dates that affect you most. For example, you would want to note any scholarship deadlines if you plan to apply.
- Create a plan. With your counselor, decide what you need to do, and make a timeline to follow. This will keep you on track with your goals.

- Make another appointment. Decide together when you will need to meet again and what you will accomplish during that meeting.
- Follow up. Complete the tasks on your timeline so you will be prepared for your next meeting.

Parents

This transitional time in your life might be one of the most difficult for both you and your parents. Probably since you were born, they've had great expectations for you. Maybe they expect you to go to a certain college or major in a certain field of study. But what if their expectations don't match your own? Although it may seem difficult, it is important to follow through with *your* own goals. If you go to college or choose a major simply to please your parents, you may find yourself unhappy and unfulfilled in a few years. There are so many decisions to make during the next year that it is important to maintain a positive relationship with your parents. While some disagreements may be expected, uncontrolled arguments and yelling at each other aren't productive. The following tips can help you and your parents communicate effectively with each other:

- Understand each other's perspective. Perhaps you've had your heart set on attending a college out of state, but your parents have told you that isn't a good option. You should find out why they feel it isn't a good option. For example, they may know they can't afford out-of-state tuition. It is often easier to work through disagreements when you understand why there is a disagreement.
- Respect each other's opinions. This doesn't mean you have to agree with everything your parents say, but you shouldn't criticize them for their opinions.
- Work to find solutions to problems. Use the problem-solving model found in this text to find a workable solution. This will keep you focused on solving the problem rather than on negative aspects of disagreeing.
- Compromise. If that out-of-state college is too expensive, perhaps you could compromise and attend college in state for the first two years. If you still want to attend the out-of-state college, you could take on the responsibility of applying for scholarships, financial aid, and student loans.
- Agree to time-outs. When disagreements become unproductive arguments, agree to let the matter rest until everyone can approach it in a positive manner.

Communicating with college representatives, high school counselors, and parents now is good practice for what you'll be dealing with on a daily basis at college.

Communicating with Others at College

The communicating and interacting you do with others while you are in college may be some of the most important communicating you will do in your life. You will need to learn to overcome fears of communicating with instructors, and you will need to learn how to effectively communicate with roommates in order to prevent and solve problems.

Instructors

First-year college students are often overwhelmed at the size of classes and the amount of work that is required. When they don't understand concepts or assignments, many students make the mistake of simply hoping they'll catch on or that another student will help them, or that they'll figure it out on their own.

However, Professor Richard L. Weaver II (who has more than thirty years of college teaching experience with more than 80,000 students) says that to be a successful student in college, one must ask questions. "If you don't ask questions, instructors assume that you know and understand, that things are fair and equitable, that you are taking advantage of opportunities, and that things are moving along properly. There is no basis for instructors to believe otherwise, unless you ask questions."[2]

Choosing not to ask questions or communicate with instructors often has a domino effect on a student: The student feels confused and afraid to admit that feeling; the student tries to muddle through without clarification or help; the student starts skipping class because he or she doesn't understand the lectures; the student has to withdraw or fail the class.

Conferencing with instructors doesn't have to be an intimidating, scary experience. Even though instructors are busy, the reason they were hired is to teach students. You should be their first priority! If you need to meet with your instructor, there are some basic ideas you should keep in mind:

- Schedule your appointment during office hours. If you absolutely can't meet during posted office hours, explain this to the instructor, and ask if special arrangements can be made. Most instructors are available at other times if you talk to them.
- When you schedule your appointment, let the instructor know why you want to meet. This will help him or her to prepare to help you.
- Be on time and be prepared. If you want to discuss a paper you've written or a test you've taken, bring a copy to the conference. Instructors grade hundreds of papers and tests throughout the course of a semester, so don't expect them to remember yours word for word.
- Have specific questions prepared and written down so you don't forget. This will help focus your time, and the meeting will be more productive.
- If there is something you don't understand, be honest. One way to check your understanding is to take notes during the conference and go over them briefly with the instructor before you leave.

After your conference, read through your notes again, and follow up with any assignments or readings. If you still are having difficulty, make an appointment with a tutor, and plan to meet with that tutor on a regular basis if necessary.

Roommates

Your college experience is often made better or worse by the relationship you have with your roommates at college. Roommates sometimes become friends for life, while other times, roommates never want to see each other again. Having honest, open communication with a roommate from the beginning can lay the foundation for a positive relationship:

- Communicate about schedules and respect each other's schedules. Perhaps one person likes to study late at night, but the other likes to go to bed early and study early in the morning. If each of you knows the other's needs, it will be easy to work out a system so that each person can be happy and productive.
- Be honest about levels of neatness. Some students don't mind clothes hung over the backs of chairs, moldy food under the bed, and empty soda cans in the closet. But living in this kind of environment can be very irritating to others. Create an agreement so that each can be happy and one isn't having to be the other person's maid.
- Discuss how you want to handle visitors. Do either of you mind having guests at all hours? Work out a plan so guests don't interfere with your academic success.
- Decide how you will deal with each other's music. If you are lucky, you'll be in agreement about music, but often music and the level that the music is played create problems. Consider buying headphones if your roommate doesn't like your music.
- Create a plan to deal with disagreements as they happen. Agree to communicate about the problem and find a mutual solution. When roommates let problems build up without dealing with them, they are setting themselves up for disaster. However, talking about the problems and learning to have balanced compromises lead to a positive living environment.

Learning to communicate with instructors and roommates is good practice for learning to communicate with supervisors on the job and even spouses. As you learn other communication strategies, your success in communicating with instructors, roommates, and employers will continue to improve.

Other Communication Success Strategies

These additional strategies can help improve your communication:

- Think before you speak. Spoken too soon, ideas can come out sounding nothing like you intended them to. Taking time to think, or even rehearsing mentally, can help you choose the best combination of words. Think it through, and get it right the first time.
- Don't withhold your message for too long. One danger of holding back is that a problem or negative feeling may become worse. Speaking promptly has two benefits: You solve the problem sooner, and you are more likely to focus on the problem at hand than to spill over into other issues.

- Communicate in a variety of ways, and be sensitive to cultural differences. Remember that words, gestures, and tones mean different things to different people.
- Be clear, precise, and to the point. Say exactly what you need to say. Link your ideas to clear examples, avoiding any extra information that can distract.

Communication is extremely important for building and maintaining personal relationships. Explore the role those relationships play in who you are.

WHY DOES GOOD WRITING MATTER?

Good writing depends on and reflects clear thinking. Therefore, a clear thought process is the best preparation for a well-written document, and a well-written document shows the reader a clear thought process. Good writing also depends on reading. The more you expose yourself to the work of other writers, the more you will develop your ability to express yourself well. Not only will you learn more words and ideas, but you will also learn about all the different ways a writer can put words together in order to express ideas. In addition, critical reading generates new ideas inside your mind, ideas you can use in your writing.

In school, almost any course you take will require you to write essays or papers in order to communicate your knowledge and thought process. In order to express yourself successfully in those essays and papers, you need good writing skills. Knowing how to write to express yourself is essential outside of school as well, as the following example demonstrates. Imagine that, after college graduation, you run a summer internship program at a major television network. You have two qualified student candidates who are vying for one internship position. Parts of both students' letters to you are shown in Figures 11.1 and 11.2.

Read both figures before continuing with this text. Which candidate would you choose? The second student's letter is well written, persuasive, logical, and error-free. In contrast, the first student's letter is not thought through clearly and has technical errors. Good writing quality gives the edge to student number two.

First student's writing sample. FIGURE 11.1

I am a capible student who'se interests are many. I like the news business so much so that I want you to offer me the internship with your company.

My experience will impress you, as I'm sure you will agree I am a reporter for the college news station, and I can be a reporter for you as well. If you will let me try. Instructers who know my work like my style. I prefer to think of myself as an individuel with a unique style that nothing can match.

FIGURE 11.2 Second student's writing sample.

From the time I was 8 years old, I was hooked on the news. Instead of watching cartoons on television, I watched Tom Brokaw, Dan Rather, and Peter Jennings. I celebrated the day that CNN started a 24-hour all-news network.

It seemed like a natural step to go into the news business. I started in high school as a reporter, and then I became editor-in-chief of the school paper. As a college freshman, I am majoring in broadcast journalism, and I am also working at the school TV station. Even though I'm starting at the bottom, I believe that there's a learning opportunity around every corner. By the time I take on my first reporting assignment next year, I feel that my knowledge and experience will have grown. I hope it will be enough to make me a competent journalist.

Instructors, admissions counselors, and other people who see your writing judge your thinking ability based on what you write and how you write it. Over the next few years, you may write papers, essays, answers to essay test questions, job application letters, resumes, business proposals and reports, memos to coworkers, and letters to customers and suppliers. Good writing skills will help you achieve the goals you set out to accomplish with each writing task.

WHAT ARE THE ELEMENTS OF EFFECTIVE WRITING?

Every writing situation is different and dependent on three elements. Your goal is to understand each element before you begin to write:

1. Your purpose. What do you want to accomplish with this particular piece of writing?
2. Your topic. What is the subject about which you will write?
3. Your audience. Who will read your writing?

Figure 11.3 shows how these elements are interdependent. As a triangle needs three points to be complete, a piece of writing needs these three elements.

Writing Purpose

Writing without having set your purpose first is like driving without deciding where you want to go. You'll get somewhere, but chances are it won't be where you needed to go. Therefore, when you write, always define what you want to accomplish before you start.

There are many different purposes for writing. However, the two purposes you will most commonly use in class work and on the job are to inform and to persuade:

The three elements of writing.

FIGURE 11.3

- The purpose of informative writing is to present and explain ideas. A research paper on how hospitals use donated blood to save lives informs readers without trying to mold opinion. The writer presents facts in an unbiased way, without introducing a particular point of view. Most textbooks and newspaper articles (except on the opinion and editorial pages) are examples of informative writing.
- Persuasive writing has the purpose of convincing readers that your point of view is correct. Often, persuasive writing seeks to change the mind of the reader. For example, as a member of the student health committee, you might write a newspaper column attempting to persuade readers to give blood. Examples of persuasive writing include newspaper editorials, business proposals, and books and magazine articles with a point of view.

Additional possible writing purposes include entertaining the reader and narrating (describing an image or event to the reader). Although most of your writing in school will inform or persuade, you may occasionally need to entertain or narrate as well. Sometimes purposes will even overlap—you might write an informative essay that entertains at the same time.

Writing Audience

In almost every case, a writer creates written material so that it can be read by others. The two partners in this process are the writer and the **audience.** Knowing who your audience is will help you communicate successfully.

Audience
The reader or readers of any piece of written material.

Key Questions About My Audience

In school, your primary audience is your instructors. For many assignments, instructors will want you to assume that they are typical readers rather than informed instructors. Writing for typical readers usually means that you should be as complete as possible in your explanations.

At times, you may write papers that intend to address informed instructors or a specific reading audience other than your instructors. In such cases,

11

you may ask yourself some or all of the following questions (depending on which are relevant to your topic):

- What are my readers' ages, cultural backgrounds, interests, and experiences?
- What are their roles? Are they instructors, students, employers, customers?
- How much do they know about my topic? Are they experts in the field or beginners?
- Are they interested, or do I have to convince them to read what I write?
- Can I expect my audience to have an open or closed mind?

After you answer the questions about your audience, take what you have discovered into consideration as you write.

My Commitment to My Audience

Your goal is to organize your ideas so that readers can follow them. Suppose, for example, you are writing an informative research paper for a nonexpert audience on using online services to get a job. One way to accomplish your goal is to first explain what these services are and the kinds of help they offer, then describe each service in detail, and finally conclude with how these services are changing job hunting in the twenty-first century.

Whether you speak or write to individuals now or when you get to college, use the strategies presented in this chapter to give you self-confidence and the power of words.

EXERCISES AND ACTIVITIES

The "I"s Have It 11.1

In your quest for better communication, rewrite the following sentences so that they are in the less accusatory "I message" style. Check your answers with other students and/or with your instructor.

1. You blew it completely.
2. Why didn't you tell me the meeting time changed?
3. You always forget to pick me up.
4. What does it take for you to understand how this machine works?
5. Where did you put the stapler? Did you lose it?
6. You are impossible to understand when you talk like that.

My Communication Style 11.2

Look back at the four styles described in this chapter: intuitor, senser, thinker, and feeler.

- Which describes you the best? Rank the four styles, listing first the one that fits best, and listing last the one that fits least.
- Of the two styles that best fit you, which one has more positive effects on your ability to communicate? What are those effects?
- Which style has more negative effects? What are they?

Endnotes

1 Sheryl McCarthy, *Why Are the Heroes Always White?* (Kansas City, MO: Andrews and McMeel, 1995), 188.

2 Richard L. Weaver II, "Ten Suggestions for Making the Most of a College Education." *Vital Speeches of the Day,* October 15, 1994, 61(1): 11–13.

MOTIVATE

MOTIVATE

12 Being a Leader

In preparing for college, applying for scholarships, getting into degree programs, and applying for jobs, one way to set yourself apart and be ready to meet rigorous program requirements is to have leadership experience and ability. Much of your course work may entail collaboration with other students on group projects, and student groups are always in need of good leaders. Leadership experience is a critical part of any academic or professional portfolio.

In this chapter, you will explore answers to the following questions:

- Why should I become a leader?
- What do leaders do?
- How can I obtain leadership skills and experience?
- How can I demonstrate and develop my leadership abilities?

WHY SHOULD I BECOME A LEADER?

If you are already a student leader or if you have observed other student leaders, you are beginning to develop a sense of the benefits of such a role. One student who saw and took advantage of leadership opportunities is Cy Salazar, a 1993 graduate of Cloudcroft High School in southern New Mexico. Throughout an academic career that included two years at New Mexico State University–Alamogordo, a branch of New Mexico State University (NMSU), and completion of a bachelor's degree at NMSU, Cy received many honors and was afforded many opportunities.

As Cy did, you will benefit from leadership experience through the contacts you will make, the communication skills you will build, and the self-confidence you will develop.

Contacts

As a leader, you make contacts that other students may not—contacts with faculty, administration, other students, and people in your community. These contacts can be valuable to you by providing the following:

- Insights about your personal mission statement
- Advice about how to reach your academic and professional goals
- Information about co-op programs, scholarships, and specialized degree programs
- Letters of recommendation
- Further contacts at colleges or in various professional fields

Cy made contacts with college administrators, alumni, legislators, and key administrators and staff at colleges other than NMSU. "The people I have worked with in my leadership roles have written letters of recommendation for scholarships, internships, and jobs," Cy said of his experiences.

Communication Skills

Leaders also develop communication skills, along with other skills such as time management, goal setting, and decision making. These are the skills that you take with you when you leave high school and college and enter the world of work.

Cy emphasizes how much his communication skills improved as his leadership career progressed. According to Cy, communications classes are excellent for building foundation skills, but "there is no substitute for being in front of real audiences with a real message to deliver. My communication skills, both oral and written, have increased with every leadership role I've been in." He continues, "I have also become better organized and better at making personal, academic, and professional decisions."

Self-Confidence

Student leaders also develop a sense of confidence and independence that other, less active students may not have until later in their careers. This confidence will show when you:

- Take on important academic projects.
- Seek higher levels of leadership responsibility.
- Interview for co-op positions or for entry into degree programs.
- Seek part-time or full-time employment.

Cy credits many of his contacts with developing his self-confidence and for offering advice and insights regarding various situations. He did not actually seek out many of the opportunities that he took advantage of; rather, the administrators and faculty who saw certain abilities in Cy guided him toward these opportunities. "A lot of what happened to me was a matter of circumstance," Cy said.

"But as I look back, I see that my leadership roles and abilities helped to create most of those circumstances." For example, Cy was nominated to serve as NMSU Student Regent even though he did not seek the position himself. He adds, "As time went on, I became more confident in myself to seek out new challenges and opportunities."

WHAT DO LEADERS DO?

According to John Gardner, there are nine tasks commonly associated with leadership roles.[1] Leaders practice these tasks at various levels, depending on the type of organization and its purpose. The nine tasks are:

1. Envisioning goals
2. Affirming values
3. Motivating
4. Managing
5. Achieving workable unity
6. Explaining
7. Serving as a symbol
8. Representing the group
9. Renewing

"Go confidently in the direction of your dreams. Live the life you have imagined."

Henry David Thoreau

Envisioning Goals

As discussed in Chapter 1, goal setting is an important beginning to any successful venture. Leaders of clubs and organizations assist the members in setting a course for the group, including short-term and long-term goals. Usually, leaders in these situations do not set goals themselves; rather, they articulate a broad vision and work with the group to set concrete goals—or steps—to attain the vision.

For example, someone leading a student organization might envision an increase in the number of active members in the organization and then work with the group to plan activities—perhaps a series of SMART goals—to bring in more members. Other goals could include the group's sponsoring certain activities on the campus or in the community, working with other groups on activities, spon-

soring a college fair or career day, or assisting the college administration in some way. Leaders who envision goals are taking steps to ensure that the group experiences lively growth and activity, not one that merely exists.

As President of the Associated Students of NMSU, Cy claims as one of his greatest successes the construction of a climbing wall on the campus. Cy and his fellow officers came to office with that goal in mind and were able to share that goal with the student community. The project included many smaller goals of obtaining funding, receiving administrative approval, planning, etc. Eventually, the project was successful, in part because of Cy's abilities to set, envision, and communicate goals.

Affirming Values

All groups—whether social, academic, religious, or political—come together because of a certain set of shared values. At the same time, individuals in the group have their own values, which differ from those of other individuals. It is the leader's task to help the group maintain focus on its shared beliefs and values. Sometimes, a leader must remind the group of its values when members drift or when conflicts among individuals threaten the health of the organization.

At times, groups can become bogged down in personality conflicts and let important work go undone. Leaders must be willing to step in and remind individuals about the shared values and purpose of the group and work to resolve any conflicts.

Two student officers of an honor society at a community college had such a conflict, becoming entangled in an ever-enlarging web of miscommunication and differing views of their responsibilities in the organization. As a result, several important projects, including the induction of new members, were delayed. The President of the society and the faculty advisors had to step in and remind the two officers of the purpose of the organization: to recognize academic achievement of students on their campus. All parties agreed to focus on the common values of the organization and to work together.

Motivating

Motivating may be among the most difficult tasks of a leader; in fact, Gardner writes: "Leaders do not normally create motivation out of thin air. They unlock or channel existing motives."[2] Often, members of a group or organization require only clear goals to be motivated to act. Affirmation of the group's common values and purpose may at times be enough motivation. Leaders can instill a can-do attitude in individuals by projecting confidence and a willingness to share in the work at hand. Sometimes, incentives or rewards can serve as motivation, but most people are motivated by the intrinsic value of work, not by external rewards.

For instance, student groups that are affiliated with regional or national organizations often compete for awards. These awards may be based on the quality and significance of projects undertaken in areas that reflect the values of the organization. For example, if a group includes community service in its mission and undertakes a service project—cleaning up a roadside, conducting a food or blanket drive, tutoring other students—participation in the project is the actual reward.

Most participants would agree that the satisfaction of completing a service project and making a difference in a community are significant motivation. This feeling of accomplishment is more valuable than any certificate, plaque, or monetary award that may result. It is the leader's task to communicate this added value of active participation in group activities, whether they are service projects, fellowship events, or academic functions.

Managing

Often, the leader of a group is the one who is accountable (the A of a SMART goal; see Chapter 1) and who must ensure that the group is progressing toward its goal. This task may include setting priorities, gathering progress reports, resolving minor and major problems within the group, and reporting back to the group. Managing also includes such basic tasks as scheduling meetings, publishing agendas, and making some executive decisions.

Leaders often delegate some of the day-to-day activities of the group: publicizing meetings, announcing special events, recording minutes, completing small tasks toward achieving a larger goal. Ideally, members of the group will be responsible and follow through. For example, those individual members who agree to solicit donations from faculty and staff for a rummage sale/fund-raiser should actually complete, in a timely manner, what they say they will do.

> "You get the best out of others when you give the best of yourself."
> *Harry Firestone*

To ensure a successful fund-raiser, however, the leader of the group should occasionally touch base with the members seeking donations and monitor progress. If progress is satisfactory, then events move forward efficiently; if progress is not satisfactory, then the leader can take appropriate steps, which may include delegating different people to the task. All communications and decisions have to be made with the ultimate goal of the project and the well-being of the organization in mind.

If this type of communication is incorporated into the group's shared values, then individual members will not feel as though they are being checked up on and will, instead, be ready and willing to report on their progress. Some of the tasks of managing may seem at times mundane, but they are critical to the success of any group or organization.

Achieving Workable Unity

Any group with more than three or four members will eventually face some conflict over goals, procedures, or financial matters. Groups may also come into conflict with other groups, particularly when they share a turf such as a school campus. No leader can ever hope to attain complete harmony within any organization; in fact, a group that finds itself without conflict will likely also find itself without growth or progress toward its goals. However, when conflict becomes an obstacle, a leader must be willing to accept some criticism from two sides of a dispute in order to guide the group back toward its shared goals and purpose. Leaders must sometimes sacrifice individual concerns or values within the organization in order to maintain the greater good—the well-being of the group itself. Again, common goals and values help achieve and maintain a workable unity by focusing individual energies within the group.

A student group may find itself divided about what projects to undertake for the year: One faction may want to sponsor a series of films and discussion sessions, focusing on a particular issue or topic, and another may want to sponsor monthly dances, cookouts, or ice cream socials. If the group's mission (yes, a student group should have a mission, as well!) is to provide students with opportunities for learning and scholarly interaction outside the classroom, then the leader must guide the "party" faction back to that shared value. If the group's mission is to provide students opportunities for fellowship and relaxation, then the leader must guide the "serious" faction in the appropriate direction.

As mentioned previously, shared values are critical to the success of any group and its activities; if this particular group does not have a shared mission or goal, then workable unity will be impossible to achieve. In this case, the task of defining the mission may fall to the leader, who must somehow achieve balance between the two factions before conflicts become destructive.

Explaining

Perhaps, communicating is a better, more inclusive term for this task. When conducting all the other tasks, leaders must be clear about the whats, the hows, and the whys of day-to-day functioning of the group or organization, as well as special circumstances. Goals must be clearly articulated and decisions clearly justified. When leaders do not explain or communicate, the group's morale can be affected, project success can be compromised, and conflicts can arise.

Cy argues that "as a leader, you have to be a good communicator; communicating is the most important thing you'll do in any leadership role." Cy credits his communication skills for his successes in running for office, in dealing with the diverse student body at NMSU, and in working with both college administration and state legislators. He said, "When I was President, I tried to make sure that the student government had at least two articles in every issue of the student newspaper; we wanted to communicate all our goals and the reasons for our decisions so that the student body knew what was going on."

Serving as a Symbol/Representing the Group

These two tasks are so similar that a joint discussion of them is necessary. Groups choose leaders in many ways: shared values, vision of the future, charisma, etc. At the core of many choices of leadership is our desire to be led by someone who represents the best of what our group values and the best of what we as individuals would like to achieve. Thus, our leaders are often symbols of the collective values, visions, and beliefs of the organization. The leader, then, *is* the organization in the eyes of the individual members.

Furthermore, no group or organization exists in a vacuum; interaction with other groups and entities is inevitable. As a symbol of the group, the leader must also then represent the group accordingly. In principle, the state and federal governments work this way; in fact, one house of a legislature is usually called the House of *Representatives*. We hope that our elected officials do indeed represent us appropriately. Leaders of student groups, in dealings with campus administration or other student groups, should not act according to their own values or beliefs, but those of the group. Any leader who abuses the trust of his or her followers and uses his or her position to push an individual agenda is acting unethically. The leader serves as the single voice of the many members and must be certain to represent them accurately, fully, and ethically.

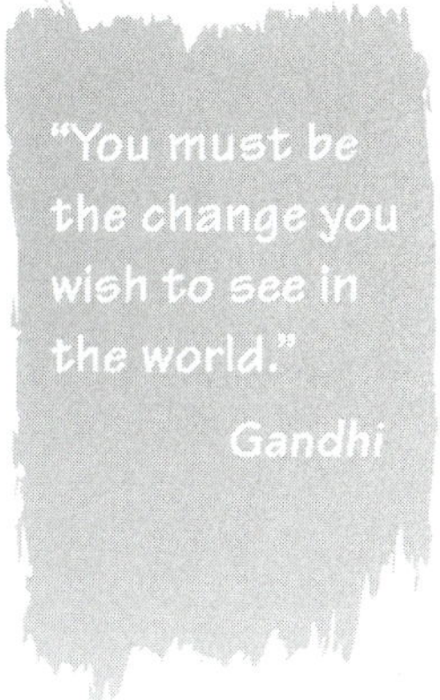

Many times, leaders serve only for the prestige or other benefits and do not take their roles seriously. Cy saw leaders of this kind damage the organizations they were associated with and eventually hurt their own leadership careers. "When I was elected President of the student body," Cy remembers, "I made an immediate effort to build relationships with the college administration; it was important that they know what was on the minds of the students and that they were able to have a direct line back to the students as well. That's the primary role of the President—to serve as liaison between the students and the college." Cy believes that this effort not only benefited him (as discussed previously) but also strengthened the relationship between the administration and the student government and student body.

Renewing

Sometimes, in addition to motivating, a leader must be responsible for renewing. Change must not be undertaken just for change's sake; however, a group should not function in a particular way just because "that's the way it's always been done." Gardner writes that any change should affirm shared values, increase vitality, and ensure the group's future.[3] All groups go through cycles of high and low activity and success. Low points often occur because leaders fail to motivate, to articulate goals, or to sense a need for renewal. Undertaking new projects, conducting membership drives, and instilling a sense of urgency or competition are ways that a leader can renew the focus and drive of a group or organization.

The previously mentioned honor society set a goal to increase the number of students who accept their invitation to membership in the organization. As part of the recruitment process, the group conducted orientations every semester; these orientations were conducted as informal come-and-go meetings during certain periods of the day and had been done this way for several years. One semester, a student officer proposed running the orientations as more formal events, with refreshments, society brochures, and an organized presentation by the officers and advisors. The other officers and members strongly resisted the change, primarily because, as one said, "That's not how it was done when I joined."

The officer justified her case by showing how the new format could present the values of the society, how it could increase the energy level and enthusiasm of the orientations, and how it could ultimately increase membership and provide for the future. After much debate, she convinced the group to try her proposal, and it proved successful. More prospective members attended the orientations, more attendees accepted membership, and more students learned about the mission of the group.

Leadership is a skill that can be learned, just like critical thinking can be learned. The best time to start learning these skills is right now. "There's nothing innate about leadership," Cy argues. "I *learned,* through experience, in classes, and from mentors, how to be a good leader."

HOW CAN I OBTAIN LEADERSHIP SKILLS AND EXPERIENCE?

When you think of leaders, you might consider your student body president, someone who holds an elected office, a movie star, or a famous team player. The people in these positions probably do exert a great deal of leadership. But you may already have some leadership experience also, which you may not recognize as such. If you are involved with student government or any other organization, keep records of projects led and completed and committees chaired; you are demonstrating leadership skills. Many schools have various clubs and organizations in which students lead and collaborate with others on fund-raising activities, academic projects, and campus events.

Cy's leadership career began with invitations to join an honor society and the community college student government association, but you should not wait for an invitation. Make leadership part of your personal mission statement and goals, and seek out opportunities. As the list of Cy's accomplishments shows, the opportunities and rewards are plentiful:

> "If you can imagine it, you can achieve it. If you can dream it, you can become it."
>
> *William Arthur Ward*

Vice President, Leadership	Phi Theta Kappa Honor Society
Recreation Chair	Student Advisory Council
Vice President	Student Advisory Council
President	Student Advisory Council
Secretary	NMSU Communications Club
Representative	Associated Students of NMSU
President	Associated Students of NMSU
Student Representative	New Mexico Commission on Higher Education
Ex Officio Member	NMSU Board of Regents

Here are some ways to begin learning leadership skills. Team sports—both organized and intramural—present many opportunities for leadership development. Everyone recognizes the leadership abilities of Michael Jordan, John Elway, and other professional figures. Team leaders perform many of the same tasks that group/organization leaders do: setting goals, motivating players, setting standards of performance, and taking responsibility. In fact, many sports leaders are effective leaders in other arenas, as well. Cy's leadership career began in high school, where he served as captain of his football and basketball teams his senior year, and intramural sports were his gateway into university life. "When I came to the university, I didn't know anyone, and intramurals gave me an opportunity to meet people and get involved in other groups and activities," he said of his transfer to NMSU.

More and more classes include collaborative learning exercises, which can also be considered sources of leadership experience: Business classes may require students to work together to create a marketing plan; instructors of computer classes may ask that students work in groups to research, design, and publish Web pages; writing classes often have students collaborating on research projects.

Students are often skeptical about participating in academic group work, primarily because their grades are determined not by their own performance but by the group's. If one student doesn't complete his or her share of the work, the grades of everyone in the group are affected. One way to avoid such problems is to assume a leadership role in any collaborative academic work. A leader of such a group would perform many of the tasks that Gardner describes.

For example, the leader might *envision goals* by having the group members decide what grade they want on the project; this would also help *motivate* the group and might help *achieve workable unity.* The leader might *affirm values* by allowing group members to work in areas of strength: One student might be a good word processor and want to finalize the printed work, and another student might be good at online research and want to collect information from the Internet. The leader might *manage* by setting deadlines for certain portions of the project and by monitoring progress, making adjustments as necessary.

The leader might *represent the group* and *explain* by meeting with the instructor to address questions and then report back to the members. The leader might *renew* and *motivate* through words of encouragement and through appreciation of work completed well and on time. *Renewal* could also be achieved by redelegating work not completed by someone not doing his or her share. The overall objective—a successful project—should always be the motivation for anything the group leader does.

In addition, leadership roles in church and community organizations should not be overlooked as sources of experience and training. If you have had such experiences and were successful in leading your group, you have experience that you can list on applications and from which you can draw to be a successful college student and leader as well.

HOW CAN I DEMONSTRATE AND DEVELOP MY LEADERSHIP ABILITIES?

At most colleges, there are numerous opportunities to continue or begin your development as a student leader. There are student groups for majors, for ethnic groups, for political interests, and for hobbies or extracurricular interests. There are honor societies, fraternities and sororities, and religious groups. At the very least, choosing two or three groups in which to become active will enable you to meet new people and make contact with people already established on the campus. These people can provide much valuable information, advice, and support as you continue to work toward your individual goals.

As you become more active, you may be assigned to chair a committee or oversee completion of a project. Usually, you'll work into leadership roles; don't expect to take over right away. Working into the role also allows you to become more familiar with the group and its values and goals; you'll also gain support from the members and be more able to perform the tasks of being a leader.

Cy's academic and leadership performance opened the door to many opportunities, including two summer internships in Washington, D.C., several job offers, and the opportunity to continue his studies in graduate school. Not everyone will be elected president of a university's student body, but everyone who seriously seeks and takes on leadership roles and responsibilities will nonetheless benefit tremendously.

EXERCISES AND ACTIVITIES

12.1 Leadership Skills in Study Groups

Break out into groups of four, and form study groups. Use leadership skills to determine goals of the groups and roles that each individual will play.

12.2 Teamwork Leadership

As a group, make a class presentation about a famous person you define as a leader. As part of your presentation, discuss how the leader performed the nine tasks discussed in this chapter.

12.3 Creating Student Organizations

Research the needs of your student body to find out if there is an organization that needs to be created. In groups, decide how to charter the organization and follow through to complete the project.

12.4 Leadership in Schoolwide Activities

As a class, sponsor an activity for the student body. Follow the tasks of leadership. When the activity is complete, evaluate the leadership tasks associated with the project.

Endnotes

1 John Gardner, "The Tasks of Leadership." In *To Lead or Not to Lead: Leadership Development Studies* (Jackson, MS: Phi Theta Kappa, 1995), 1.11–1.17.

2 Ibid.

3 Ibid.

Index